Bibliografische Information der Deutschen Nationalbibliothek

Die Deutsche Nationalbibliothek verzeichnet diese Publikation in der
Deutschen Nationalbibliografie; detaillierte bibliografische Daten sind
im Internet über http://dnb.d-nb.de abrufbar.

ISBN 978-3-8325-2766-2

Logos Verlag Berlin GmbH
Comeniushof, Gubener Str. 47,
10243 Berlin
Tel.: +49 (0)30 42 85 10 90
Fax: +49 (0)30 42 85 10 92
INTERNET: http://www.logos-verlag.de

Design of a Pulsed Frequency Modulated Ultra-Wideband System for High Precision Local Positioning

Entwurf eines gepulsten frequenzmodulierten
Ultrabreitband-Systems zur hochgenauen
Positionsbestimmung

Der Technischen Fakultät der
Universität Erlangen-Nürnberg
zur Erlangung des Grades
DOKTOR-INGENIEUR

vorgelegt von
Benjamin Waldmann

Erlangen – 2011

Als Dissertation genehmigt von
der Technischen Fakultät der
Universität Erlangen-Nürnberg

Tag der Einreichung: 27. Oktober 2010
Tag der Promotion: 13. Dezember 2010
Dekan: Prof. Dr.-Ing. habil. Reinhard German
1. Berichterstatter: Prof. Dr.-Ing. Dr.-Ing. habil. Robert Weigel
2. Berichterstatter: Prof. Dr.-Ing. Martin Vossiek

Acknowledgements

This work was written during my work as a research assistant at the Institute for Electronics Engineering of the Friedrich-Alexander University Erlangen-Nuremberg. Most importantly, I would like to thank Prof. Dr.-Ing. Dr.-Ing. habil. Robert Weigel, who has given me the opportunity to work in his institute and to do a doctorate. Despite his many duties and obligations he always keeps an eye on his doctoral candidates and follows their work with constant interest and is always open to their problems and needs.

I am also very grateful to Prof. Dr.-Ing. Martin Vossiek for taking the position of a reviewer on my doctorate committee and his support and advice throughout the whole UPOS research project. He had the first idea, together with Dr.-Ing. Peter Gulden, for the PFM-UWB technique.

Many thanks go also to the company *Symeo GmbH*, especially to Dr.-Ing Peter Gulden and Dipl.-Ing. Dirk Becker for their constant advice and support. Their invaluable technical knowledge, their willingness to help and the always open door in Munich contributed significantly to the success of this work. A demonstrator would have never come to this state without their help. I would also like to thank especially Dr.-Ing. Sven Roehr, who sacrificed whole weekends to help me. Without his aid it would have been hard to finish this work, especially the measurement chapter. THANK YOU!

I also thank all colleagues at the Institute for Electronics Engineering in Erlangen for the excellent work atmosphere. In particular, I thank Dr.-Ing. Ralf Mosshammer, Dr.-Ing. Kay Uwe Seemann, Dr.-Ing. Marcus Hartmann, Dr.-Ing. Henning Ehm, Dr.-Ing. Errikos Lourandakis, Dr.-Ing. Benjamin Sewiolo, Dipl.-Ing. Christoph Kandziora, Dipl.-Ing. Jochen Essel, M.Sc. Gabor Vinci, and Dipl.-Ing Stefan Zorn. Through many fruitful discussions, they helped me out of some blind alleys and were thereby estimated companions, and comrades in misfortune. Numerous exhilarating evenings created friendships that will last over the times as a doctoral student at the LTE.

I would also like to mention my students that helped with their studies and theses to built up a real UWB demonstrator. In particular I like to highlight Dipl.-Ing. Alexander Goetz and wish him all the best for his scientific future.

Finally, I thank my family and my wife Anja for their support and recourse, especially during the last half year in which most of this work was written, in particular on weekends and in the late evening hours. They kept me grounded and helped to motivate when no end was in sight. Therefore this work is dedicated to them.

The every day challenge to give my best, to create something new, to be creative and to change the world, for this I get up each morning.
For this and for Gutmann Hefeweizen.

When you finish reading this book,
tie a stone to it and throw it into the
Euphrates.

Jeremiah, 51, 63

Abstract

Over the last decades an ever growing demand for wireless localization systems for many different indoor and outdoor application scenarios could be observed. Applications range from crane and fork-lift positioning to RFID-like asset tracking and access control and many more. Especially indoor applications show severe performance degradation due to heavy multipath effects. Ultra-Wideband technology opens up new fields of microwave applications for communications and ranging and positioning. Particularly for the latter application field, the full potential of UWB is not yet used efficiently. Since the resolution of wireless localization systems for all modulation techniques is proportional to the applied bandwidth, systems using ultra-wideband signals offer an excellent potential for accurate indoor positioning. With the accredited bandwidth for ultra-wideband (UWB) systems, an accuracy in the range of a few centimeters is possible.

This thesis presents the analysis and design of a pulsed frequency modulated ultra-wideband system for high precision local positioning. Up to now, all commercial available systems are based on the usage of very narrow pulses to achieve ultra-wideband signals. To overcome typical limitations of pulsed UWB technology, a novel system concept is presented in this work. It combines pulse technology with frequency modulated continuous wave (FMCW) radar technology by chopping the FMCW signal into short pulses. On the one hand this fulfills the UWB regulation requirements, and on the other hand precise control of the signal shape is given for the whole bandwidth. This allows the use of advanced synchronization and distance measurement approaches.

The main emphasis of this work is the investigation and realization of this pulsed frequency modulated (PFM)-UWB positioning system. This comprises a close study of the UWB indoor channel to derive the requirements on the developed positioning system. It is shown that especially in industrial environments, a large signal bandwidth is mandatory to achieve a good overall system performance. The PFM-UWB system is based on a well-known LPR system, which is extended to an UWB system. Theoretical calculations, simulations and measurements are used to define the system and evaluate its performance. A prototype was built which operates around the center frequency of 7.5 GHz utilizing a bandwidth of 1 GHz. With this demonstrator, a high accuracy in dense multipath indoor environments can be achieved. In normal operation mode, the achievable standard deviation during distance measurements is well below 1 cm and the absolute error in an office indoor environment is below 3 cm in 85% of all measurement cases, which makes the system ideally suited for 2D and 3D indoor localization. Conclusive measurements, that demonstrate the abilities of the developed positioning system, round off this work. The results of the measurement campaign show the excellent performance of the proposed system concept, and demonstrate the significant advancement compared to the state of the art.

Zusammenfassung

In den letzten Jahrzehnten konnte eine stetig wachsende Nachfrage nach drahtlosen Lokalisierungssystemen für den Einsatz in verschiedensten Bereichen beobachtet werden. Die Anwendungen reichen dabei von der Kran- und Gabelstapler-Ortung, RFID-ähnlicher Warennachverfolgung bis hin zu Zutrittskontrollsystemen und vielen mehr. Vor allem bei Anwendungen für den Innenbereich zeigen existierende Systeme dramatische Leistungseinbußen aufgrund der starken Mehrwegeausbreitung der Funksignale. Hier bietet die Ultrabreitband-Technologie neue Möglichkeiten der Funktechnik für Kommunikations- und Lokalisierungssysteme. Gerade für letztere Anwendungen wird allerdings das volle Potenzial der Ultrabreitband-Technik noch nicht effizient genutzt. Da das Auflösungsvermögen drahtloser Lokalisierungssysteme generell proportional zur genutzten Signalbandbreite ist, bieten Ultrabreitband-Systeme die Möglichkeit einer hochgenauen Ortung von Personen und Objekten innerhalb von Gebäuden. Eine Genauigkeit im Bereich von wenigen Zentimetern ist mit dieser Technik auch unter schwierigen Bedingungen möglich.

Im Zentrum der vorliegenden Arbeit steht die Analyse und das Design eines gepulsten, frequenzmodulierten, ultrabreitbandigen Systems zur hochpräzisen Positionierung und Ortung. Bis zum Zeitpunkt der Erstellung dieser Arbeit basierten sämtliche kommerziell erhältliche UWB-Funkortungssysteme auf der Nutzung sehr schmaler Pulse. Um die typischen Nachteile dieser gepulsten, ultrabreitbandigen Technologien zu überwinden, wird ein neuartiges Systemkonzept im Rahmen dieser Arbeit vorgestellt. Durch das Zerhacken eines linear frequenzmodulierten Signals verbindet das hier vorgestellte System die klassische Pulstechnik mit der bekannten FMCW-Radar-Technik. Mit Hilfe dieser Technik kann das System, in Übereinstimmung mit den verschiedenen internationalen Regulierungsbehörden, als UWB-System klassifiziert werden und erfüllt gleichzeitig die entsprechenden Anforderungen und Limitierungen.

Den Schwerpunkt dieser Arbeit bildet die Untersuchung und der Entwurf eines vollständigen, gepulsten, frequenzmodulierten, ultrabreitbandigen Lokalisierungssystems. Dieses umfasst eine genaue Analyse des UWB-Indoor-Funkkanals, um daraus die Anforderungen an das Ortungssystem abzuleiten. Es wird gezeigt, dass insbesondere in industrieller Umgebung eine große Signalbandbreite zwingend notwendig ist, um eine hohe Genauigkeit des Systems zu erreichen. Das vorgestellte System basiert auf einem bekannten LPR-Prinzip, welches zu einem UWB-System erweitert wird. Mit Hilfe von theoretischen Berechnungen, Simulationen und Messungen wurde das System definiert und voruntersucht und anschließend ein Prototyp realisiert. Dieser nutzt eine Mittenfrequenz von 7,5 GHz und eine effektive Signalbandbreite von 1 GHz. Mit diesem Demonstratorsystem ist es möglich, eine hohe Lokalisierungsgenauigkeit - auch bei starker Mehrwegeausbreitung - zu erzielen. Unter verschiedenen Betriebsbedingungen liegt die erreichbare Standardabweichung bei Abstandsmessungen deutlich unter 1 cm und der absolute Fehler bei Messungen innerhalb eines Bürogebäudes in 85% der Fälle unterhalb von 3 cm. Das System ist somit ideal für die Realisierung von 2D- und 3D-Lokalisierungsanwendungen, speziell für den Indoor-Bereich, geeignet. Zum Abschluss der Arbeit werden Messergebnisse präsentiert, welche das Leistungsvermögen des entwickelten Ortungssystems demonstrieren. Die Ergebnisse der Messkampagne zeigen die hervorragende Leistungsfähigkeit des vorgeschlagenen Systemkonzepts und stellen damit einen wesentlichen Fortschritt gegenüber dem aktuellen Stand der Technik dar.

Contents

i

List of Figures

List of Tables

Abbreviation	Description
ADC	analog-to-digital converter
AOA	angle-of-arrival
AWGN	additive white Gaussian noise
CDMA	code devision multiple access
CEPT	European Conference of Postal and Telecommunications Administrations
CIR	channel impulse response
CW	continuous wave
DDS	direct digital synthesizer
DFT	discrete Fourier transform
DS-CDMA	direct-sequence code devision multiple access
DSP	digital signal processor
DUT	device under test
EC	European Commission
ECB	equivalent complex baseband
ECC	Electronic Communications Committee
EIRP	equivalent isotropically radiated power
ETSI	European Telecommunications Standards Institute
FCC	Federal Communications Commission
FD	frequency domain
FDMA	frequency-division multiple access
FFT	fast Fourier transform
FMCW	frequency modulated continuous wave
GPS	Global Positioning System
IEEE	Institute of Electrical and Electronics Engineers
IF	intermediate frequency
IFFT	inverse fast Fourier transform
IMU	inertial measurement unit
IR	impulse response
ISM	industrial, scientific and medical
LNA	low-noise amplifier
LO	local oscillator
LOS	line-of-sight
LPR	local positioning radar
MPC	multipath component

Abbreviation   Description

NLOS	none-line-of-sight
OFDM	orthogonal frequency-division multiplex
PCB	printed circuit board
PDF	probability density function
PDP	power delay profile
PFD	phase frequency detector
PFM	pulsed frequency modulated
PG	path gain
PLL	phase-locked loop
PMD	photonic mixer device
PSD	power spectral density
RBW	resolution bandwidth
RCS	radar cross section
RF	radio frequency
RFID	radio frequency identification
RMS	root mean square
RSC	Radio Spectrum Committee
RSS	received signal strength
RTLS	real-time location system
RTOF	round-trip time-of-flight
S-V	Saleh-Valenzuela
SNR	signal-to-noise ratio
SPDT	single-pole double-throw
SRD	step recovery diode
TD	time domain
TDMA	time-division multiple access
TDOA	time-difference-of-arrival
TOA	time-of-arrival
TOF	time-of-flight
UPos	ultra-wideband positioning system for complex environments
UWB	ultra-wideband
VBW	video bandwidth
VCO	voltage controlled oscillator
VNA	vector network analyzer

CHAPTER 1

Introduction

1.1 Motivation

Over the last decades, wireless communications underwent an enormous growth and became an integral part of our daily lives. Wireless technology range from global satellite communications systems to wireless local area networks: a diversity of technologies, enabling us to stay connected anytime and anywhere. One of the major challenges for future wireless systems is to deliver an increased performance like a higher data rate and at the same time not to disturb existing systems. UWB is a promising technology to fulfill these two requirements. UWB technology is not new but became its major push by the release of the Federal Communications Commission (FCC) part 15 frequency regulations in 2002 [1], which allowed the unlicensed use of UWB devices under certain emission constraints. Due to the low transmit power, UWB devices can coexist with existing wireless technologies and even share their frequency spectrum. Furthermore it uses very large signal bandwidths, which allow for high data rates in communications systems.

Another important benefit of the usage of very large signal bandwidths is the ability to realize very high accurate and precise ranging and positioning systems. According to the popular Internet encyclopedia *Wikipedia*, positioning systems are defined as "a mechanism for determining the location of an object in space. Technologies for this task exist ranging from worldwide coverage with meter accuracy to workspace coverage with sub-millimeter accuracy" [2]. The accuracy requirements for such positioning systems strongly depend on the application the system is designed for. Table 1.1 gives a short overview over some application fields and their accuracy requirements according to [3]. The global positioning system (GPS) is probably the most common positioning system and many location-based services rely on this prominent system.

Applications	Accuracy
Advertising	100[m]
Train / air / bus information	30[m]
Local information	30[m]
Docking	5[m]
Public services tracking	3[m]
Location-based services	3[m]
Access control	3[m]
Precision landing	1[m]
In-building worker tracking	1[m]
Pedestrian route guidance	1[m]
Goods and item tracking	1[m]
Exhibit commentary	1[m]
Incidence tracking / guidance	80[cm]
Recreation and toys	10[cm]
In-building robot guidance	8[cm]
Tool positioning	1[cm]
In-building survey	1[cm]
Automated handling	0.5[cm]

Table 1.1: Accuracy requirements for different application examples.

But since GPS and other satellite based positioning systems are not able to deliver reliable data for indoor positions, alternative solutions have to be found. Especially in the industrial field there is a growing interest on high accurate real-time location systems (RTLS) for the monitoring and controlling of complex production processes. Fig. 1.1 shows the global market on real-time location system (RTLS) from 1998 to 2005 after [4], where an exponential growth can be observed. According to *Idtechex's* forecast for the global RTLS market for the year 2016 it will increase from 70 million U.S. dollars in 2006 to a total of 2.7 billion U.S. dollars [4]. This market can be divided into three major sectors: military, health care, and industrial and logistics as shown in Fig. 1.2. This increasing interest and enormous growth underlines the significance and importance of localization systems. In general there is not *one* localization technology, that fits for all application scenarios. But the unique aspects of UWB make it a very promising technique to realize diverse kinds of ranging and positioning systems, especially in the very harsh industrial environment.

The objective of this work was to develop a tool tracking system for the use at industrial production lines. According to various quality requirements during a production process in the automotive industries (e.g. according to ISO/TS 16949 or QS 9000) it is necessary to generate a quality assessment for every manufacturing step. At the same time, the production has to be most efficient and flexible. To ensure this requirement, mobile tools and parallel, independent work are necessary on several pieces. This high flexibility and mobility during the production process is in contrast to the requirements on quality assurance. Therefore the overall objective of this research work was to develop a solution for quality monitoring of production steps, in which mobile tools are involved. Thus allowing to better connect the different requirements on quality, effectiveness, and flexibility with each other than has been possible before. An elementary basic functionality for monitoring and quality assurance of mobile tools is the automatic real-time tracking of the tools. It is an elementary function, that allows to examine which production step at which position and time was made and whether steps were omitted. To solve this very challenging task, a new UWB microwave radio tracking system was developed for tool positioning. The proposed solution goes far beyond the currently available state of technology and provides a significant innovation in the field of industrial positioning.

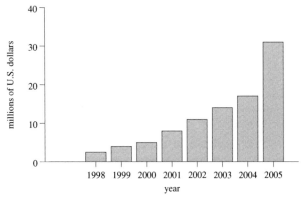

Figure 1.1: Development of the global market on RTLS from 1998 to 2005 according to [4].

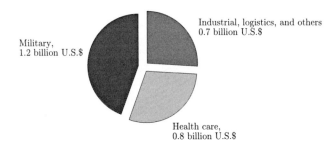

Figure 1.2: Major applications of future RTLS in 2016 according to [4].

1.2   Ultra-Wideband Localization Systems: State of the Art

As mentioned before, there is a large bandwidth of available positioning systems on the market. This section summarizes the most significant UWB positioning systems as existing at the beginning of this work in 2006. The usage of UWB signals for ranging and positioning systems has its breakthrough in the year 2002 with the approval of the unlicensed use of the 3.1 GHz to 10.6 GHz frequency spectrum by the FCC. A selection of the most prominent UWB positioning systems are listed below:

Ubisense

Ubisense [5] was founded in 2002 and has its roots at the Laboratory of Communications Engineering at the University of Cambridge. The company provides a fully developed real-time localization system and achieves a 3D accuracy better than 15 cm in indoor environments according to their publications [6–8]. Fig. 1.3 shows the base station, called "Sensors", and two available tag variants "Slim Tag" and "Compact Tag". The tags transmit an UWB impulse signal in the frequency range of 6 GHz to 8 GHz. The sensors are equipped with an antenna array and estimate the angle-of-arrival (AOA) of the incoming signal. In addition to that, the system is also able to operate in a time-difference-of-arrival (TDOA) mode for which the sensors have to be connected via a "timing-cable" for synchronization purposes. The tags are triggered and controlled via an additional wireless link in the 2.4 GHz industrial, scientific and medical (ISM) band. The maximum measurement rate is 160 Hz per tag.

Figure 1.3: Hardware of the *Ubisense* UWB positioning system. The tags, available in two variants "Slim Tag" (b) and "Compact Tag" (c), are worn by a person or attached to an asset and transmit impulse signals, received by the "Sensor" (a). The "Sensor" is equipped with an array of four UWB receivers, allowing an AOA estimation. By connecting the sensors via a "timing-cable" the system is able to operate in a TDOA mode. With these combined techniques the position of the tags is estimated.

Precision Location Ultra-Wideband System (PLUS)

The Precision Location Ultra-Wideband System (*PLUS*) was developed by *Time Domain* [9] and is also based on impulse technology. In 1999, *Time Domain* were the first to introduce ultra-wideband radios using full custom integrated circuits [10, 11]. A full analog UWB correlation receiver was designed for communications purposes, which also allowed an round-trip time-of-flight (RTOF) distance measurement. After several steps of further developments, *Time Domain* now offers the *Plus* system, which locates tags based on TDOA measurements. For this purpose short UWB impulses are used. The system is available in different regional variants, where the center frequency is either 6.6 GHz or 7.3 GHz, depending on the the local UWB regulations. The complete system is comprised of tags, readers, and synchronization distribution panels (SDP). The readers and the SDP are connected with a shielded cable, so that a common time base can be distributed very precisely. In Fig. 1.4 the *PLUS* product line consisting of two available antennas, a 6-port synchronization distribution panel, a reader and two tag variants "Badge Tag" and "Asset Tag" is shown. *Time Domain* specifies the achievable indoor accuracy within sub-meter range and a maximum update rate of 25 Hz [12].

Figure 1.4: Hardware of the *Time Domain* Precision Location Ultra-Wideband system (*PLUS*). The system consists of synchronization distribution panels (a), distributing an accurate clock amongst the receivers (b). Tags are available in two variants "Badge Tag" (c) and "Asset Tag" (d) and transmit impulse signals, which are received by the readers (b). The position estimation is performed via TDOA. Additionally *Time Domain* offers a set of antennas (d) optimized for their system.

Sapphire DART Ultra-Wideband

The *Sapphire DART Ultra-Wideband Localization System* was developed by *Multispectral Solutions (MSSI)*, now part of *Zebra Enterprise Solutions (ZES)* [13]. The system is based on the works of J.R Fontana [14–16] and also uses TDOA measurements for the position estimation of various tags. The hardware setup is comprised of UWB receivers, which are connected and synchronized over a dedicated cabling infrastructure, transmit-only tags, that are localized, and a central processing hub. The tags, available in the variants "1x1 Asset Tag", "Mini-Badge Tag", and "Micro Asset Tag", transmit short pulses in the frequency range of 5.94 GHz to 7.12 GHz. According to their publications [17–19], the system achieves an accuracy of better than 30 cm over a maximum operating range of 45 m. The maximum measurement rate is given with 25 Hz.

Others

Besides these three commercial available UWB localization systems there are many research and university projects which focus on the development of high accuracy UWB ranging and positioning systems.

In [20] a non-coherent UWB system with positioning capability is presented. Synchronization and time-of-arrival (TOA) estimation is performed using a non-coherent energy collection method. According to the publications, a sub-meter accuracy was achieved with this method.

The German IMST demonstrated in [21] an impulse based AOA system for indoor localization. With this system, an angular accuracy of 1 ° to 4 ° was achievable, leading to a positioning uncertainty of 10 cm to 20 cm.

A research group at the Japanese company Fujitsu presented a high accurate positioning system in 2007 [22]. The system uses short UWB impulses and an energy detection-based receiver architecture. The radio frequency (RF) signal is located in the 3.7 GHz to 5 GHz frequency range. According to their publications, the system is able to achieve an accuracy below 20 cm and an indoor operating range of 8 m.

An outstanding performance was reported by a research group at the University of Tennessee [23–25]. They demonstrated an architecture for UWB positioning systems, which combines the architectures of carrier-based UWB systems and traditional energy detection-based UWB systems using a sub-sampling technology at the receiver side. With this technique, a 3D accuracy of a few millimeter in a TDOA setup was reported. However, the operating range is limited to a few meters.

Overall, there is a very wide range of various research groups and commercial systems utilizing UWB technologies for high-accuracy indoor positioning. Obviously impulse technology is up to now the predominant technique for UWB localization systems. Most of these systems use the TDOA approach for the position estimation and need a very precise synchronized backbone network, which is realized via a cable infrastructure. The achievable accuracy, as well as the measurement rate and operating range vary widely depending on the system and the applications the systems are designed for. Table 1.2 provides a summary of the different systems and their key parameters.

Company / Research Group	Frequency (GHz)	Accuracy / Reported Error	Operating Range
Ubisense	6 - 8	<15 cm	90 m
Zebra Enterprise Solutions	5.94 - 7.12	<30 cm	45 m
Time Domain	$f_c = 6.6$, or $= 7.3$	<1 m	not reported
Cheong et al.	not reported	<1 m	not reported
IMST	not reported	<20 cm (2D)	not reported
Fujii et al.	3.7 - 5	20 cm (2D)	8 m
Zhang et al.	5.4 - 10.6	5 mm	5 m

Table 1.2: Comparison of current research and commercial UWB positioning systems.

1.3 Goals and Organization of this Work

At the time the thesis was started, available UWB localization systems did not use the full potential of the technique. Especially for the afore mentioned application scenario of a tool tracking system in industrial environments none of the available positioning solutions could fully satisfy the requirements. Within a research project UPos (Ultra-Wideband Positioning for Complex Environments) alternative UWB techniques were investigated and a demonstrator system was developed. Before the system could be developed, the environment in which the system should operate had to be investigated. From this fact the first goal is derived: to investigate and model the UWB channel in an typical industrial environment. This work includes an intensive investigation of UWB channel modeling and the main differences to conventional narrowband channel models. By gaining information about the channel, the technical requirements on a positioning system can be derived. The second goal was to develop an UWB based localization system using PFM signals and implement a first hardware demonstrator to show the potentials of this new and highly innovative technique.

After the introduction, the next chapter focuses on the definition of UWB signals and the international regulations for UWB signal emissions. Chapter 3 gives a short overview of the principles of local positioning systems. The chapter presents position estimation and tracking techniques, which are applicable to any signaling technique and not limited to UWB signals. UWB channel modeling techniques are discussed in chapter 4, including differences between narrowband and UWB channel models. Also the results of a channel measurement campaign at an industrial fabrication hall are presented. In chapter 5 the developed pulsed frequency modulated UWB technique and its application in the local positioning system is presented. First a short overview over the *Symeo* local positioning system (LPR) is given, since it is the base for the newly developed localization system. The section is followed by a short highlighting of multipath issues and how they can be solved using very large signal bandwidths. Afterwards the expansion of the conventional system towards the PFM-UWB system and the UWB-local positioning radar (LPR) is shown and how it is realized in a hardware demonstrator. Chapter 6 gives some measurement results and a performance analysis of the local positioning system. In chapter 7 the thesis concludes by summarizing the results and giving an outlook for future work.

CHAPTER 2

Ultra-Wideband Signals

In this chapter an overview of the development of the ultra-wideband technology is given along with a description of the principles and the most important parameters and characteristics of this technology. For a more detailed introduction into the history and the basics of UWB it is referred to [26–28]. Here only the most important facts are repeated and summarized which are essential for the understanding of the later described UWB positioning system.

The emerging technology of UWB has a long and changing history throughout the past years. The predominant method of wireless signal transmission today is based on sinusoidal waves. This technique has become so universal in radio communications that many people are not aware that the first communication systems were in fact pulse-based. In 1887 it was the German physicist Heinrich Hertz who used a spark discharge to produce electromagnetic waves for his experiments [29]. Spark gaps and arc discharges between carbon electrodes were the primary wave generators for about 20 years after Hertz's first experiments. In the following years sinusoidal signal forms became the dominant form of wireless microwave systems and it was not until the 1960s that work began again in earnest for time domain electromagnetics. The development of the sampling oscilloscope in the early 1960s and the corresponding techniques for generating sub-nanosecond baseband pulses sped up the development of UWB. Impulse measurement techniques were used to characterize the transient behavior of certain microwave networks.

The main focus moved from measurement techniques to radar and communication devices. Early nomenclatures of UWB technology include *baseband, carrier-free, non-sinusoidal* and *impulse* and the field of UWB had moved in a new direction. Other applications, such as automobile collision avoidance, positioning systems, liquid-level sensing and altimetry were developed. At the beginning it was mainly used by military forces for radar, sensing and secure communication. Especially radar was given a lot of attention because of the accurate results that could be obtained and the possibility to penetrate objects or the ground due to the low-frequency components of baseband pulses. Therefore

9

it was the U.S. Department of Defense who established the term UWB in the late 1980s. The late 1990s witnessed a move towards the commercialization of UWB communication devices and systems. Start-ups like *Time Domain, Xtreme Spectrum* or *Multispectral Solutions* were formed to profit from this development. In February 2002 the Federal Communications Commission (FCC) issued the rule for using UWB in a band of 7.5 GHz [1]. Never before such an enormous bandwidth had been allocated for a wireless and terrestrial system offering potentially data rates up to 1 Gigabits per second. Another important fact is that the band for UWB is license free. To put that into perspective, the auctions of the licenses for UMTS generated a total sum of 100 billions Dollar for the European governments . These are the two main reasons why the industry is working to establish UWB for the mass market.

2.1   Definition of Ultra-Wideband

UWB signals are characterized by their very large signal bandwidth compared to the conventional narrowband systems. In general a signal is called UWB if it has an absolute bandwidth of at least 500 GHz, or a fractional (relative) bandwidth larger than 0.2. The absolute bandwidth B_{UWB} is defined by the difference between the upper frequency f_h and the lower frequency f_l, defined in Fig. 2.1:

$$B_{UWB} = f_h - f_l. \tag{2.1}$$

The frequencies f_l and f_h mark the -10 dB emission points. Therefore B_{UWB} is also called the -10 dB bandwidth. The fractional bandwidth is defined by

$$B_{frac} = \frac{B_{UWB}}{f_c}, \tag{2.2}$$

where f_c is the center frequency and is given by

$$f_c = \frac{f_h + f_l}{2}. \tag{2.3}$$

 UWB systems are characterized by very short duration waveforms, usually in the order of a nanosecond. Commonly, an UWB system transmits ultra short pulses with low duty cycle. In other words, the ratio between the pulse transmission instant and the average time between two consecutive transmissions is usually kept small. This type of UWB systems that transmits UWB pulses with low duty cycle is called impulse radio (IR). In an IR UWB communications system, a number of pulses are transmitted per information symbol and information is usually conveyed by the positions or the polarities of the pulses. A detailed description of IR UWB communications systems can be found in [27, 30]. Besides for communications systems, the impulse technology is the dominant technique in UWB radar and positioning systems.
Another possible implementation of UWB systems is the use of continuous transmitted signals. For example, a Direct Sequence Spread Spectrum system (DS-CDMA) with a very short chip interval can also be used as an UWB communication system [31]. In addition to that, the transmission and reception of very short orthogonal frequency-division

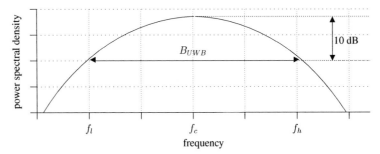

Figure 2.1: An UWB signal is defined to have an absolute bandwidth $B_{UWB} \geq 500\,\mathrm{MHz}$, or a fractional bandwidth of $B_{frac} = B_{UWB}/f_c > 0.2$.

multiplex (OFDM) symbols can also be considered as an OFDM UWB scheme [30]. However, for UWB communications and radar systems, both low duty cycle schemes and continuous transmissions can be considered.

The common property of the usage of a very large bandwidth holds many advantages for communications, radar and positioning systems. These advantages can be summarized as follows:

• Penetration through obstacles

• High ranging and positioning accuracy

• High-speed data communications

• Low cost and low power implementation

The large frequency spectrum of UWB signals, including low frequencies as well as high frequencies, provide the possibility of material penetration and result in a high time resolution. Therefore UWB is an ideal technique for all kinds of positioning, imaging and radar applications.

2.2   International Regulation Issues for UWB Systems

As stated before, applying UWB techniques is a very promising concept for both communications and localization systems. However, the usage of very large bandwidths of several GHz requires UWB systems to work as an overlay of already allocated frequency bands. Therefore UWB systems must not influence or disturb already established services in the occupied frequency range. For example, frequency allocation of some wireless systems is shown in Fig. 2.2. Thus UWB transmitters must meet certain requirements in order to avoid any negative effects on other systems. In February 2002, the Federal Communications Commission (FCC) issued the FCC UWB rules that have provided the first radiation limitations for UWB and have permitted the commercialization of this technology. According to these regulations, the maximum permitted radiated power of UWB systems is strictly limited to very low levels. Due to the low power density of the transmitted signal, the UWB signal is similar to noise and is not likely to cause significant interference to existing radio systems in the same frequency spectrum.

After legalizing the usage of UWB systems in the USA by the "First Report and Order", the development and standardization efforts were pushed worldwide. In the following, the FCC regulations are investigated in detail. And to this the regulation status in Europe at the time of writing this thesis is examined closely.

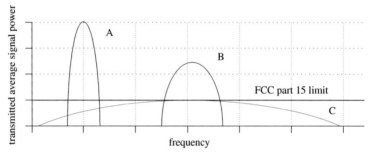

Figure 2.2: Spectral distribution of different wireless systems. A. Conventional narrowband system, B. Spread spectrum system, C. UWB. Note that the bandwidths and power levels are not drawn to scale.

2.2.1   UWB Regulations in the USA

In the USA, UWB technology was released under the FCC Part 15 regulation limitations and allowed a license-free use of UWB products. The Part 15 regulations for UWB systems impose certain power emission limits for various types of UWB systems. Within these rules intentional radiators are permitted to operate in certain frequency bands. They are not permitted to operate in sensitive or safety-related frequency bands like the frequency band of the GPS system. The FCC defines an UWB transmitter as an intentional radiator that, at any point in time, has a fractional bandwidth equal to or greater than 0.2 or has a UWB bandwidth equal to or greater than 500 MHz, regardless of the fractional bandwidth. This excludes explicitly systems, that use stepped or swept frequency techniques to fulfill the minimum bandwidth requirement, which is an important issue, discussed in chapter 5. The FCC defines seven types of UWB systems:

- Ground penetrating radars and wall imaging systems

- Through-wall imaging systems

- Surveillance systems

- Medical imaging systems

- Vehicular radar systems

- Indoor UWB systems

- Hand held UWB systems

The emission limits for UWB systems are specified in terms of equivalent isotropically radiated power (EIRP), that is defined as the product of the power supplied to the antenna and the antenna gain ($g_{ant|dB}$) in a given direction relative to an isotropic antenna as given in Eq. 2.4.

$$EIRP = 10^{\frac{g_{ant|dB}}{10}} P. \tag{2.4}$$

The maximum EIRP of UWB emitters must not exceed a maximum of -41.3 dBm/MHz, which is equal to the limit for unintentional radiators, such as televisions or computer monitors. Depending on the specific application area some systems must have even lower limits in some frequency bands. According to the FCC regulations, emissions are to be measured using a resolution bandwidth of 1 MHz. A detailed description of the appropriate measurement technique is given in appendix A. Since this work presents a novel positioning system for indoor environments the FCC limits for indoor UWB systems are given below. A detailed description of the requirements of all types of systems can be found in [1].

Indoor UWB devices have to be designed for indoor use only and the emissions from the transmitters must not be intentionally directed outside of a building such as through a window or a doorway. Furthermore, the occupied UWB bandwidth has to be between 3100 MHz and 10600 MHz. Fig. 2.3 shows the corresponding emission mask while the average limits of the radiated emissions from a device are given in table 2.1. In addition to this maximum allowed radiated average power, there is a limit of 0 dBm EIRP on the peak level of the emissions contained within a 50 MHz bandwidth centered on the frequency at which the highest radiated emission occurs.

Frequency [MHz]	EIRP [dBm/MHz]
960-1610	-75.3
1610-1990	-53.3
1990-3100	-51.3
3100-10600	-41.3
Above 10600	-51.3

Table 2.1: FCC emission limits for indoor UWB systems.

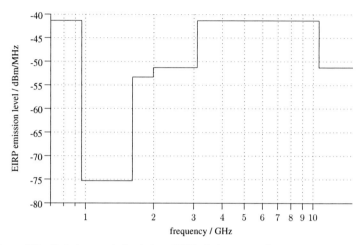

Figure 2.3: Emission mask for indoor UWB devices according to the FCC regulations. For the frequency range from 3.1 GHz to 10.6 GHz a maximum EIRP density of -41.3 dBm/MHz is allowed.

2.2.2   European Regulations

After the release of the FCC regulations for UWB systems, regulatory efforts have been made worldwide. Although a consistent regulation with worldwide acceptance would be desirable, the emission limits are different in various parts of the world. In Europe, the Electronic Communications Committee (ECC) of the European Conference of Postal and Telecommunications Administrations (CEPT) undertook several technical studies for UWB regulations, which were considered by the Radio Spectrum Committee (RSC) of the European Commission (EC). Eventually at the beginning of 2007 the EC made the final decision for UWB regulations, that are valid in the member countries including Germany [32]. The EC has chosen to make use of only part of the spectrum that was approved for use in the U.S. in 2002. The European decision defines UWB technology as technology involving the intentional generation and transmission of RF energy that spreads over a frequency range wider than 50 MHz. Thus making no restrictions towards stepped or swept frequency systems like the Part 15 definition, which is an important fact for the PFM-UWB positioning system. In Fig. 2.4 the spectrum mask for UWB systems is shown, that do not employ appropriate interference mitigation techniques. For these systems a frequency range of 6 GHz to 8.5 GHz is available with a maximum EIRP of -41.2 dBm/MHz. At the moment the frequency range between 4.2 GHz to 4.8 GHz is only provisionally available until the end of 2010. From 2011 the according EIRP will be limited to -70 dBm/MHz for that band. In other bands there are more restrictive limits compared to the Part 15 regulations.

Systems that use interference mitigation techniques are allowed to transmit -41.3 dBm/MHz in the 3.4-4.8 GHz band. Fig. 2.5 shows the corresponding emission mask with the according limits summarized in table 2.3. Systems operating in the lower frequency band have to provide a low duty cycle transmission. This is specified in terms of T_{on} and T_{off}, which are defined as the duration of a burst and the duration between two consecutive bursts. This burst duration should not exceed 5 ms ($T_{on} < 5$ ms) and the total off-time should be larger than 950 ms per second whereas the total on-time should not exceed 5% per second and 0.5% per hour.

Frequency [MHz]	EIRP [dBm/MHz]
<1610	-90
1610-3400	-85
3400-3800	-85
3800-4200	-70
4200-4800	-41.3 (until 31 December 2010)
	-70 (beyond 31 December 2010)
4800-6000	-70
6000-85000	-41.3
8500-10600	-65
Above 10600	-85

Table 2.2: Maximum EIRP densities in absence of appropriate mitigation techniques.

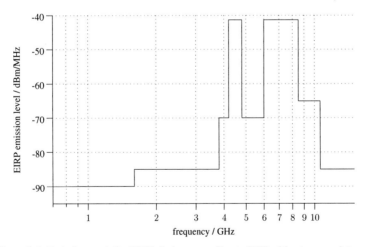

Figure 2.4: Emission mask for UWB devices according to ECC without appropriate mitigation technique. Until 2011 UWB systems are allowed to transmit -41.3 dBm/MHz in the 4.2-4.8 GHz band, after then, the emission limit is lowered to -70 dBm/MHz.

Frequency [MHz]	EIRP [dBm/MHz]
<1610	-90
1610-2700	-85
2700-3400	-70
3400-4800	-41.3
4800-6000	-70
6000-8500	-41.3
8500-10600	-65
Above 10600	-85

Table 2.3: ECC emission limits with appropriate mitigation technique.

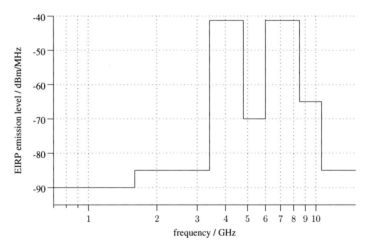

Figure 2.5: Emission mask for UWB devices according to ECC with appropriate mitigation technique.

CHAPTER 3

Principles of Local Positioning Systems

The knowledge of the accurate position of a certain target is a more and more important factor for all kinds of operation fields. Examples for such applications are tool tracking, indoor guidance, security applications, surgery assistance, and many more. The requirements for these localization system strongly depend on the aimed application. The GPS is suited for outdoor scenarios with moderate accuracy requirements. With this technique it is possible to reach an accuracy in the meter range but it suffers from high accuracy degradation in multipath environments and, as mentioned before, it is normally limited to outdoor applications. To overcome these limitations a couple of alternative solutions were introduced, which will be presented in the following. In general there are four main technologies available for wireless localization: optical/infrared, ultrasound, inertial sensors, and radio frequency systems.

The optical systems offer a very broad range of techniques. The most important ones like the *Optotrack* system by *Northern Digital Inc.* [33] or the *Firefly* by *Cybernet Systems Corporation* [34] rely on the detection of active infrared markers using high resolution cameras. With these kinds of systems it is possible to achieve a sub-mm accuracy and very high measurement rates. Further optical systems are visual based like the 3D time-of-flight (TOF) photonic mixer device (PMD) technique [35]. The main drawback of optical systems is that they are sensitive to direct sunlight and rely on a good line-of-sight condition. Furthermore the typical operation range is below 10 m.

Ultrasound systems are an attractive solution for wireless localization in harsh environments. The comparatively slow propagation velocity of ultrasound allows for a precise distance and position measurement. Examples are the *Sonitor* system by *Sonitor Technologies* [36] or the *Cricket* indoor location system [37]. Enabling a very high accurate 3D position estimation, ultrasound systems suffer from a limited operation range of approximately 10 m and are very sensitive to environmental factors such as other ultrasound sources.

A third possibility to estimate the position and movement of targets, is the use of inertial sensors like the foot-mounted inertial sensor system presented in [38]. These sensors measure the acceleration and angular rate in order to calculate the movement and the position of a target. In general, three angular rate sensors (gyroscopes) and three acceleration sensors (accelerometers), which are orthogonally positioned to each other, are required for this purpose. With these systems a very high accuracy can be achieved. The major drawback of those systems is the need for a re-calibration after a certain time since the errors of the sensors accumulate over time, making a pure inertial sensor localization system impossible.

In this chapter the fourth method using radio frequency signals for ranging and positioning, is highlighted. Over the years different techniques and principles emerged for radio based positioning. Depending on accuracy requirements and constraints on transceiver design, various RF signal parameters can be employed to estimate the position of a target. In general three measurement categories can be distinguished:

- The usage of the received signal strength (RSS)

- The estimation of the angle-of-arrival (AOA)

- Time-of-flight (TOF) based systems, like time-of-arrival (TOA), time-difference-of-arrival (TDOA) and round-trip time-of-flight (RTOF) systems

In the following these principles are introduced followed by a short discussion on the figures of merit for localization systems.

3.1   Received-Signal-Strength Principle

The measurement of the received signal strength is a simple and inexpensive approach for localization systems. The power, or energy, of a signal transmitted by a node is a signal parameter that contains information related to the distance from the transmitter. In theory there is a linear relationship between the received signal strength and the distance between the transmitter and the receiver. Therefore the distance can be determined by mapping the path loss of the received signal to the distance traveled by the RF signal. However this obvious use of the RSS is usually not reliable since especially in indoor environments there are several more effects besides the path-loss, which lead to significant variations of the RSS like for example obstructions and multipath effects. To overcome this unstableness, the position estimation is in general performed through the use of a method known as fingerprinting. A generalized architecture is shown in Fig. 3.1. In such an approach, calibration measurements have to be performed at certain reference points to generate a RSS map of the environment. The measured values of the receiver are then mapped on this calibration data and the real position is obtained by database correlation/pattern-matching algorithms [39,40]. Therefore, algorithms for RSS positioning systems usually implement training phases to obtain a mapping function from the database. Although generic algorithms are employed to improve the position estimation, RSS localization has limited applicability in unknown or changing environments. Especially in harsh environments with a distinctive multipath characteristic, the performance of RSS based systems degrades rapidly. Nevertheless a coarse position estimation can be achieved with an accuracy in the range of 2 m to 10 m. The possibility to reuse existing wireless local area network infrastructure allow for an easy integration of RSS localization systems in an existing RF infrastructure [41]. Therefore several commercial available systems like the *Real-Time Location System* by *Ekahau* [42] emerged. For further information and detailed discussions on RSS based localization systems, see [43–47].

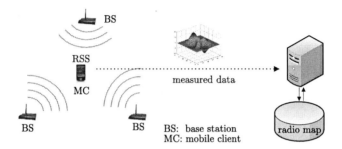

Figure 3.1: A generalized illustration of a received-signal-strength based positioning system. The mobile client measures the received-signal-strength and sends these values to a central processing unit, where the measured parameter are compared with a predefined radio map. The RSS value is mapped to a certain point in space and the position of the mobile is calculated.

3.2   Angle-of-Arrival Principle

Another method to obtain the position of a mobile unit is the usage of the direction of an incoming signal or the angle-of-arrival. For this method, the base stations are commonly equipped with an antenna array in order to measure the AOA. The angle information is obtained by estimating the phase difference of the incoming signal at the different antenna elements. The principle of this method is illustrated in Fig. 3.2 using an uniform linear antenna array. Given a sufficient large distance between the transmitter and the receiver with the antenna array, the incoming signal can be modeled as a planar wave-front. Depending on the angle-of-arrival, the RF signal arrives the different antenna elements with a certain time delay given with $\sin\left(\Psi/c_0\right) l$, where l is the inter-element spacing of the antenna array. For conventional narrowband systems, this time delay transfers into a phase difference $\Delta\Phi$ between the channels. By estimating this phase difference, the AOA can be obtained. For wideband systems, time delayed versions of received signals should be considered, since a time delay cannot be represented by a unique phase value for a wideband signal [48]. More advanced array structures, such as uniform circular arrays and rectangular lattices, operate on the same basic principle [48–51]. In Fig. 3.3 the principle setup of an AOA localization system is depicted, where multiple base stations detect the direction of an emitted signal. In general two base stations are enough to achieve an accurate position estimation using simple geometric rules. Each estimated angle gives a line between the mobile unit and the corresponding base station. The intersection of these lines provide the position of the mobile. The accuracy of an AOA system diminishes with an increasing distance due to the scattering environment. Another drawback of this measurement principle is the need for line-of-sight (LOS) conditions and the need for extra hardware, especially for very sophisticated antenna architectures, since the performance of the positioning system strongly depends on the accuracy of the used antennas. Especially in the field of radar techniques, several algorithms have been developed to evaluate multichannel signals for angle-of-arrival estimations [52–54]. However, sole angle-of-arrival based positioning systems are not a very suitable technique in indoor environments.

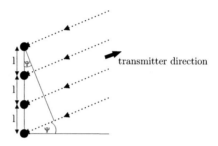

Figure 3.2: Angle-of-arrival measurement at a uniform linear antenna array.

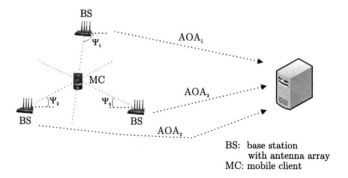

Figure 3.3: A generalized illustration of an AOA based positioning system. The base stations are equipped with antenna arrays, which allow the estimation of the angle-of-arrival of an incoming signal. The intersection of at least two vectors gives a unique position of the mobile.

3.3   Time-of-Flight Systems

To obtain a very high accurate position determination, it is necessary to measure distances via the time-of-flight of a signal that travels between two wireless nodes, for the direct timing information provides the best way to calculate distances and hence the position of a certain unit. The use of time-of-flight of RF signals to measure distances and obtain positioning informations is not a new concept. The most popular system of that kind is the GPS which uses the one-way delay of radio waves transmitted by satellites to estimate the distance. The propagation-time based systems can further be subdivided into three measurement principles, which will be described in the following: time-of-arrival, time-difference-of-arrival, and round-trip time-of-flight.

3.3.1   Time-of-Arrival Principle

Time-of-arrival measurements are performed to obtain information about the distance between two units by estimating the time-of-flight of an RF signal that travels from one unit to another. For this measurement principle it is necessary that the mobile, as well as the reference units all have very precisely synchronized clocks. The reference stations estimate the time-of-arrival of the incoming microwave signal transmitted by a mobile unit. Since the point in time when the signal is transmitted is known, the time-of-flight of the signal and hence the according distances to the individual reference stations can be calculated. Fig. 3.4 shows an exemplary setup of a time-of-arrival system. Here, all the involved nodes are synchronized via a backbone network. At the time t_0, the mobile unit transmits a signal, which is received at the ith base station after the time-of-flight τ_i. Assuming that there is an ideal LOS radio channel, with this time τ_i, the distances d_i between the mobile client and the reference stations can be calculated using the speed of light:

$$d_i = c_0 \tau_i. \tag{3.1}$$

The TOA measurements provide an uncertainty region in the shape of a circle with the reference station as the center point. By measuring the distances to several reference stations a 2D or 3D localizing can be obtained by estimating the interception point of these circles. The absolute achievable accuracy of a TOA positioning system is depending on the effective bandwidth of the used RF signal. Generally speaking, the higher the bandwidth, the better the distance estimation. Since UWB signals are characterized as signals with a very large bandwidth, this property allows for highly accurate position estimation using TOA measurements [55].

The main drawback of this localization principle is the need for a very high clock accuracy, since errors in the synchronization of the units correspond directly to errors in the position estimation. If for example a position accuracy in the decimeter range is aimed for, an absolute time synchronization significantly below 1 ns is needed for all involved nodes [56]. This imposes significant hardware costs, especially on the mobile client side.

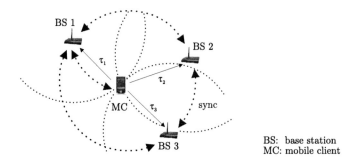

BS: base station
MC: mobile client

Figure 3.4: A generalized illustration of a time-of-arrival based positioning system. All the nodes including the reference stations, as well as the mobile client have an highly synchronous time base. At a certain point in time t_0 the mobile transmits a signal, which is received by the base stations at the times-of-arrival t_i. With this knowledge, the times-of-flight τ_i and hence the distances between the base stations and the mobile client can be calculated. The position is obtained by triangulation.

3.3.2   Time-Difference-of-Arrival Principle

The usage of the time-difference-of-arrival is a very similar technique to that of TOA , but in this case, the difference between the distances from a mobile unit to each base station is calculated. The most obvious advantage of the TDOA measurement principle is that it is only necessary to synchronize the reference nodes [56]. This can be done either by a wired infrastructure or a wireless reference transponder located at a well-known location [57]. Having a very precisely synchronized backbone network, it is possible to measure the difference between the arrival times of a a signal received by several reference nodes and transmitted by the mobile unit. Considering an exemplary setup of one mobile client located at the coordinates x,y and a set of i synchronized reference stations as shown in Fig. 3.5. Each base station depicts the time-of-arrival of a received signal and obtains a timing value t_i consisting of the absolute time-of-flight τ_i and an unknown timing offset t_0 due to the unknown transmitting point in time. The distance from the mobile to the ith base station at the location x_i, y_i can easily be expressed by using the Pythagorean theorem

$$\sqrt{(x - x_i)^2 + (y - y_i)^2} = c_0 \left(\tau_i - t_0 \right), \tag{3.2}$$

with the two unknown position coordinates x and y. The difference in the distance between two reference stations and the mobile unit is obtained by subtracting the corresponding distance equations, which results in a hyperbolic function given with

$$\Delta d_{ij} = c_0 \left(t_i - t_j \right) \sqrt{(x - x_i)^2 + (y - y_i)^2} - \sqrt{(x - x_j)^2 + (y - y_j)^2}. \tag{3.3}$$

Solving the equation locates the mobile on a hyperbola with the two reference stations as the focal points. The particular branch of the hyperbola that the target is on is the one that is closest to the reference station that receives the signal first. Since this one time difference value is not enough to calculate the two coordinates x and y, TDOA requires one more reference unit compared to the TOA principle. By determining the intersection of at least two hyperbola, it is possible to estimate the two-dimensional coordinates of the mobile unit. Like TOA systems, the accuracy of TDOA positioning systems also increases with bandwidth, which again makes UWB signals very attractive.

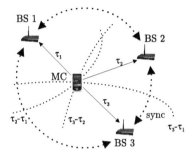

BS: base station
MC: mobile client

Figure 3.5: A generalized illustration of a time-difference-of-arrival based positioning system. The base stations are synchronized via a backbone network. At an unknown time, the mobile client transmits a signal, which is received by the reference units. Since these have a common time base, the difference of the arrival times at the individual base stations can be calculated and therefore the difference of the corresponding distances. This technique locates the mobile on hyperbolas and by estimating the intersection point, the position of the client can be obtained. For a 2D position estimation, a minimum of three reference stations is needed.

3.3.3   Round-Trip Time-of-Flight

The RTOF measurement technique is a third way of estimating a 1D distance between a reference station and a mobile unit. The measurement principle is closely related to the well-known common radar principle and illustrated in Fig. 3.6. In this simple approach, a unit transmits an RF signal, which arrives at the target after the time-of-flight τ. From there, it is reflected to the transceiver and received after twice the time-of-flight $t = 2\tau$. With the knowledge of this round-trip time, the distance between the unit and the target can easily be calculated via

$$d = \frac{c_0 t}{2}. \qquad (3.4)$$

Compared to the previously described TOA and TDOA systems, an RTOF system completely overcomes the problem of having a highly synchronized infrastructure.

Figure 3.6: The principal function of a simple passive backscatter system. A transceiver emits an RF signal, which is reflected by the target object. After the round-trip time-of-flight 2τ, the signal is received by the measuring unit. With this time the distance between the transceiver and the reflector can be calculated.

Instead of simply reflecting the interrogating radar signal, a more advanced responding unit can reflect the signal coherently superimposed with a certain modulation scheme [58]. Again there is no need for a synchronization, but the measuring unit has to know the exact delay caused by the signal processing at the responder. This modulated backscattering allows for a distinction of the wanted signal and signals which are reflected by obstacles or other reflectors and received by the measuring unit. Backscatter systems are very common in a broad field of applications and are often combined with communications systems like radio frequency identification (RFID) [59]. A major drawback of this system is that the signal has to travel the complete path between the transmitter and the responder twice. The propagation loss is therefore proportional to the fourth power of the distance between the units d. By using an active transceiver as the responding unit, this problem can be overcome. Furthermore a more complex secondary unit allows for more sophisticated modulation and multiple access schemes like time-division multiple access (TDMA) or frequency-division multiple access (FDMA). In these kinds of systems, an RF signal is transmitted by the measuring unit, received by the transponder, which synchronizes on the incoming signal and sends a synchronized and modulated reply back to the measurement unit. Depending on the properties of the RF signal it is possible to achieve very precise distance measurements with an accuracy in the range of a few mm. Examples of such powerful RTOF systems are presented in [60]. In Fig. 3.7 an exemplary setup of such an RTOF positioning system is shown. Here, the base stations act as the measuring units and transmit an RF signal. This signal is received by the mobile client which sends a synchronized response back to the reference units. From the round-trip time-of-flight, the distance of the mobile unit to the individual base stations can be calculated. The mobile is located on circles with the known position of the reference units as the center points. The multidimensional position is gained by calculating the intersection of at least three circles similar to the position estimation in TOA systems. To distinguish the responses of the different reference stations, a multiplex scheme, like TDMA or FDMA has to be adopted. It is also possible to use the mobile client as the measuring unit and the base stations as responder. In this case the position estimation is done by the mobile client.

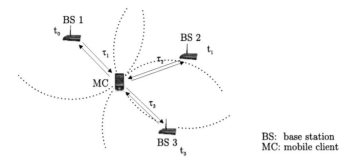

Figure 3.7: A generalized illustration of a round-trip time-of-flight based positioning system. Here the base stations transmit a signal, which is received by the mobile client. After synchronization, the mobile sends a response back to the reference unit, which depicts the distance using the round-trip time-of-flight. In this example TDMA is used to differentiate the individual reference units. The position estimation is done via three time slots t_0, t_1, and t_2, where the distances are estimated consecutively. The position of the mobile is obtained by triangulation.

3.4   Other Measurement Types

In numerous multidimensional positioning systems a combination of position-related parameters are estimated and combined instead of performing a single measurement. With the fusion of several parameters, a more accurate position can be obtained compared to a system that uses only a single position parameter. In such hybrid systems various combinations of measurement techniques are feasible. A common method is to combine TOA and AOA measurements [61–63]. The main principle of this position estimation technique is shown in Fig. 3.8. The TOA measurement principle locates the mobile on a circle with the center at the position of the base station while an AOA measurement provides a direction of the transmitter. The intersection of the circle and the direction vector gives the unique position of the mobile.

Furthermore various combinations of measurement techniques, such as TOA/RSS [64,65] or TDOA/AOA [66,67] are possible, depending on accuracy requirements and complexity constraints.

In addition to these techniques it is also possible to measure the multipath power delay profile (PDP) or the channel impulse response (CIR) of a certain environment which allows for a position estimation via a fingerprinting method [68–70]. Depending on the used RF signals, a PDP or CIR measurement can contain significantly more information about the position of a certain unit compared to the other measurement principles. To obtain the actual position from these kind of measurements, a method similar to the described RSS fingerprinting and mapping has to be adopted. Systems equipped with an antenna array can also employ multipath angular power delay profile measurements.

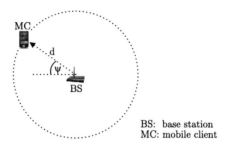

BS: base station
MC: mobile client

Figure 3.8: A generalized illustration of a hybrid TOA/AOA system. Via a time-of-flight measurement the distance between the base station and the mobile unit is obtained. The measurement of the angle-of-arrival gives the direction vector. The intersection between the circle and the vector provides a unique position estimate for the mobile using only one reference station.

3.5   Tracking

Using the techniques described in the previous sections, the position of a node is estimated, based on a single measurement. This value gives only a more or less rough position estimation, depending on the accuracy of the used positioning technique. The single position estimation is affected by error mechanisms like Gaussian noise for example. Therefore, in positioning systems, the position is estimated via the mean value over multiple raw measurement data. The accuracy of the position estimations can be improved by using the position history of the node for the final position estimation. In recent years Kalman and Particle filters became the dominant techniques for the implementation of tracking filters. The Kalman filter produces estimates for the true value of position measurements and their associated calculated values by predicting a value, estimating the uncertainty of the predicted value, and computing a weighted average of the predicted value and the measured value [71]. One of the main advantages of using Kalman filters is its computational efficiency in obtaining the optimal solution.

Particle filters, also known as sequential Monte Carlo methods, represent the posterior probability of the state at a given time by a set of samples, called particles, and related weights, called importance factors [72]. A major drawback of particle filtering is its computational complexity, which increases exponentially with the number of particles.

The two methods can also be combined by using a version of the Kalman filter as a proposal distribution for the Particle filter.

Kalman and Particle filters have been employed in numerous positioning and tracking systems. A detailed description is beyond the scope of this work, a few examples are given in [73–77].

3.6 Accuracy versus Precision

Two figures of merit of positioning systems are precision and accuracy. The dictionary definitions of these two words do not clearly make the distinction as it is used in the science of measurement. For an evaluation of a positioning system, the distinction of these two parameters is essential.

Accuracy in this context describes the deviation of a measured value in respect to its real value. It is a measure of rightness. It is given by the difference of a measured and real distance between two nodes for example. The accuracy is mainly effected by physical parameters of a positioning system like the radio channel. A main parameter for determining the accuracy of a ranging and positioning system is the resolution, which gives the capability to distinguish two different targets for classical radar applications [78] or the smallest position difference the system can detect, in positioning systems. In general, the resolution and hence the accuracy of a ranging and positioning system is indirect proportional to the applied bandwidth of the system. If the distance is estimated from a number of measurements, the precision can be defined as the deviation of a single measurement from the mean value over all measurements, supposing a quasi-static scenario. It is therefore a measure of exactness or repeatability. The precision is affected by noisy, nonlinearity, and other nonideality effects of the measurement system itself. Fig. 3.9 shows two examples for a high accurate (a) and a high precision system (b). The measurement results are Gaussian distributed for both cases. The precision can therefore be described by the standard deviation of the distribution of the measured values and the accuracy can be derived from the difference of the mean value from the real x-value.

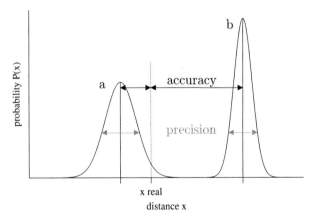

Figure 3.9: Comparison of an accurate (a) and a precise (b) ranging system. While the accuracy depicts the deviation from the real position, the precision expresses the repeatability of single measurements. For a Gaussian distribution, the precision can be defined as the standard deviation and the accuracy as the difference of the mean value from the real value.

CHAPTER 4

Ultra-Wideband Radio Channel

A proper channel modeling has significant importance for gaining insights into designing physical layer systems and selecting certain system parameters. For UWB systems, the envisaged channel is very similar to a wideband channel as may be experienced in common spread spectrum or CDMA systems. The main distinguishing feature of an UWB channel is the extremely high time resolution of the channel impulse response. This property allows to distinguish reflections caused by surfaces separated by mere centimeters. Various channel modeling techniques like ray tracing and statistical modeling have been studied extensively in literature. In this chapter only the recently proposed UWB channel models are presented and their interpretations for positioning applications according to [26]. First the differences between UWB and narrowband systems are explained with the according consequences for channel modeling. Afterwards the development of an UWB channel model is explained, starting with the simple Saleh-Valenzuela model which is the basis for the standardized channel model according to the IEEE working group 802.15.4a. At the end of the chapter a simple channel sounding method is introduced with measurement results obtained in a typical industrial environment.

4.1 UWB versus Narrowband

The extreme large bandwidths of UWB systems lead to some significant differences in modeling of the radio channel compared to common narrowband or wideband systems. In latter systems, the properties of objects, such as their reflections and scattering parameters can be modeled as frequency independent constants due to the small used frequency bands. This assumption is not valid for UWB systems . In the following, the main differences between UWB and narrowband channel characterizations are discussed focusing on two examples. The first is an ideal free-space propagation scenario, the second is a more realistic propagation environment with multipath effects.

4.1.1   Free-Space Propagation Channel

In Fig. 4.1 a very simple free-space propagation scenario is shown. The signal transmitted by the transmitter reaches the receiver only via the direct path over a distance of d. The frequency-dependence of the free-space propagation can be found in any textbook on wireless communications. The path gain G_{PG} at a certain frequency f and propagation distance d is given as

$$G_{PG}(d, f) = \frac{P_{RX}}{P_{TX}} = G_{TX}(f)\, \eta_{TX}(f)\, G_{RX}(f)\, \eta_{RX}(f) \left(\frac{c_0}{4\pi f d} \right)^2, \qquad (4.1)$$

where G_{TX} and G_{RX} are the antenna gains for the transmit and the receive antenna respectively, and η_{TX} and η_{RX} are the antenna efficiencies. c_0 stands for the speed of light. Equation 4.1 shows that if the gains of the antennas vary significantly with frequency, the path gain changes as well. In contrast to narrowband systems, where the path gain can be considered as frequency independent, this gain can vary significantly over frequency in UWB systems, which has to be accounted for. One possibility of modeling the path gain for UWB radio channels will be given in section 4.2. The antenna efficiency η is mostly determined by the matching and is also frequency dependent. A factor depending on the matching is the impedance bandwidth, which specifies a frequency band over which the signal loss is not very significant. Designing good UWB antennas with sufficient matching over a wide frequency range is a very challenging task, details on this topic can be found in [27, 30].

Figure 4.1: Transmitter and receiver constellation with free-space signal propagation.

4.1.2   Reflection and Transmission

In a more realistic environment, where obstacles and walls are present, the signal propagation becomes more complex. Fig. 4.2 shows a simple drawing of signal propagation with multipath effects. The transmitted signal arrives at the receiver via the direct path with the length d and in addition to that, it is reflected by an object and reaching the receiver over the paths d_1 - d_2. Therefore, the properties of the objects in an environment are also very important in determining the characteristics of a radio channel. The path gain G_{PG} for the reflected signal arriving at the receiver can be expressed as

$$G_{PG}\,(d_1, d_2, f) = G_{TX}\,(f)\ \eta_{TX}\,(f)\ G_{RX}\,(f)\ \eta_{RX}\,(f)\ \frac{c_0^2 \sigma_{RCS}\,(f)}{4\pi\,(4\pi f d_1 d_2)^2}. \qquad (4.2)$$

σ_{RCS} represents the radar cross section (RCS) of the object. It can be considered as a fictional surface area that intercepts the incident wave and scatters the energy isotropically in space. In narrowband systems, the RCS is modeled as a constant, however for UWB systems the RCS must be considered as a frequency-dependent parameter, which can lead to significant fluctuations in the received signal power over frequency.

A second effect has to be considered when comparing narrowband and UWB channels. The transmission through obstacles, especially for ground or wall penetrating systems must be considered as strongly frequency-dependent. The transmission through a dielectric layer of width d_{layer} is given by

$$T = \frac{T_1 T_2 e^{-j\alpha(f)}}{1 + \rho_1 \rho_2 e^{-2j\alpha(f)}}, \tag{4.3}$$

where T and ρ are the transmission and reflection coefficients. Index 1 denotes air, and index 2 denotes the considered material. The quantity $\alpha(f)$ is the frequency-dependent electrical length of the dielectric as seen by waves that are at an angle Θ_t with the layer

$$\alpha = \frac{2\pi}{c_0} f \sqrt{\epsilon_{r,2}}\, d_{layer} \cos(\Theta_t), \tag{4.4}$$

where $\epsilon_{r,2}$ is the relative dielectric constant of the layer material.

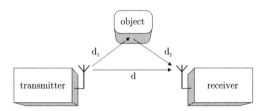

Figure 4.2: Transmitter and receiver constellation with one reflective object.

4.1.3 Others

Besides the aforementioned effects, there are several others that show strong frequency dependence. Diffraction at the edge of a screen or wedge, or scattering on rough surfaces are two examples for that. Since a detailed discussion of these effects would be beyond the scope of this work, further information and detailed discussions can be found in [79].

4.2   UWB Channel Modeling

For the simulation and testing of wireless systems, two methods to characterize UWB channels are common. The first approach is called deterministic modeling. In this method, it is assumed that complete geometric information and electromagnetic characteristics of the environment is available. With this knowledge, an electromagnetic simulation tool with ray-tracing techniques is used to determine the signal propagation characteristics in that environment. This approach leads to a very accurate representation of the radio channel, with the drawback of being site-specific. If anything changes in the geometric environment, the model can easily get invalid for the new environment constellation. Furthermore the gathering of sufficient and accurate informations of the environment can get very complex and cumbersome and the simulation very time and computer resource consuming.

A second way of modeling a radio channel is the so-called statistical-modeling approach. In that approach, statistical models are derived from actual channel measurement campaigns and are less complex than the deterministic modeling. These stochastic channel models reflect the essential properties of propagation channels, without trying to emulate the exact behavior for each specific location. The key parameters that represent the behavior of the radio channel and therefore have to be modeled accurately are path gain, large-scale fading, also referred to as shadowing, power delay profile, and small scale fading. The modeling of these parameters for an UWB positioning system are discussed in the following.

4.2.1   Path Gain

Path gain (PG) is defined as the ratio of the received signal power to the transmitted signal power, and for UWB systems it is a frequency-dependent parameter as discussed in the previous section. When the received signal power shows fluctuations due to multipath or shadowing, the path gain for a narrowband system at a distance d is conventionally defined as

$$G_{PG}(d) = \frac{E\left\{P_{RX}(d, f_c)\right\}}{P_{TX}}, \tag{4.5}$$

with f_c as the center frequency. The expectation $E\left\{\right\}$ is taken over an area that is large enough to allow averaging of the shadowing as well as the small scale fading. Due to the discussed frequency-dependency the UWB path gain (PG) becomes a function of distance as well as of frequency and can be expressed as

$$G_{PG}(d, f) = E\left\{\int_{f-\Delta f/2}^{f+\Delta f/2} \left| H\left(\tilde{f}, d\right) \right|^2 d\tilde{f}\right\}, \tag{4.6}$$

where $H(f, d)$ is the channel transfer function, and Δf is chosen small so that the material properties can be considered as constant within that bandwidth. The total path gain is obtained by integrating over the whole bandwidth of interest. According to [80] the path gain can be simplified by dividing it into a function of the distance and a function of the frequency and therefore can be rewritten as

$$G_{PG}(d, f) = G_{PG}(f) G_{PG}(d). \tag{4.7}$$

The distance dependency of the path gain for UWB systems is the same as in most used narrowband channel models. There are many investigations in literature on this issue, which can be reused here. The distance dependent path gain is described by the well known expression

$$G_{PG}(d) = G_{PG,0} - 10n \log_{10} \left(\frac{d}{d_0} \right). \qquad (4.8)$$

Usually the reference distance d_0 is set to 1 m, $G_{PG,0}$ is the path gain at the reference distance and n is the propagation exponent. This exponent strongly depends on the environment and the transmit conditions e.g. whether there is a line-of-sight or not. The exponents for an indoor environment vary in literature from 1.0 in a narrow corridor, to 1.2 in industrial environments, to $1.5 - 2.0$ in office and residential environments [81–83]. For non-line-of-sight scenarios typical exponents range from 2 to 7. Since the focus of this work lies on LOS scenarios for detailed discussions on NLOS propagation exponents it is referred to [81, 84–88].

In [89, 90] the frequency dependent part of the path gain is given as

$$G_{PG}(f) \propto f^{-2\kappa}, \qquad (4.9)$$

where κ denotes the frequency decaying factor varying between -1.4 and $+1.5$ depending on the environment and whether the antenna effects are included or not [82, 91]. As a result of the frequency dependence of the path gain, the well known wide-sense stationary uncorrelated scattering (WSSUS) assumption [92] is not valid for UWB channel modeling.

4.2.2   Large-Scale Fading

Large-scale fading also referred to as shadowing, depicts the variation of the mean signal power around the path gain. It is commonly modeled as a lognormal distribution with a typical variance between 1-6 dB depending on the environment and transmit scenario [81, 84, 86, 87, 93]. By adopting this premise, the path gain of eq. 4.8 becomes

$$G_{PG}(d) = G_{PG,0} - 10n \log_{10} \left(\frac{d}{d_0} \right) - S, \qquad (4.10)$$

where S is a Gaussian-distributed random variable with zero-mean and standard deviation σ_{sh} [80].

4.2.3   Channel Impulse Response

The channel impulse response (IR) gives the delay dispersion of a certain channel. The channel impulse response usually consists of the first peak, which corresponds to the line-of-sight signal path, followed by multiple multipath component (MPC). For the UWB channel it had been recognized in many investigations, that these MPCs tend to arrive in clusters. The most common way to model this effect is the use of the Saleh-Valenzuela

(S-V) model. It was developed by Adel Saleh and Reinaldo Valenzuela in 1987 and was proposed in [94] to model the multipath effects of an indoor environment for wideband channels, with a bandwidth in the order of 100 MHz. Even at this relatively narrow bandwidth, a clustering phenomenon was observed in the channel. In order to capture this effect, Saleh and Valenzuela proposed an approach that modeled the multipath arrival times using a statistically random process based on the Poisson point process. Additionally, the impulse response of the channel is divided into clusters, which consist of several rays. The impulse response in equivalent complex baseband (ECB) representation can be expressed as

$$h(t) = \sum_{l=0}^{\infty} \sum_{k=0}^{\infty} a_{k,l} e^{j\phi_{k,l}} \delta(t - T_l - \tau_{k,l}), \tag{4.11}$$

where T_l depicts the arrival time of the lth cluster and $\tau_{k,l}$ the arrival time of the kth ray relative to the arrival time of the lth cluster, with $k, l \geq 0$ as well as $T_0 = 0$ and $\tau_{0,l} = 0$. The amplitude of every ray is given by $a_{k,l}$ and the according phase by $\Phi_{k,l}$. The time interval between two consecutive arrival times follows an exponential distribution

$$p(T_l \| T_{l-1}) = \Lambda e^{-\Lambda(T_l - T_{l-1})} \tag{4.12}$$

$$p(\tau_{k,l} \| \tau_{k-1,l}) = \lambda e^{-\lambda(\tau_{k,l} - \tau_{k-1,l})}, \tag{4.13}$$

with Λ and λ representing the arrival rates. The amplitude coefficients $a_{k,l}$ are random, but the root mean square value depends on the arrival time of the cluster and the ray and can be expressed as

$$\overline{a_{k,l}^2} = \overline{a^2(T_l, \tau_{k,l})} = \overline{a^2(0,0)} e^{-T_l/\Gamma} e^{\tau_{k,l}/\gamma}. \tag{4.14}$$

The probability density function of the amplitude coefficients is Rayleigh distributed:

$$p(a_{k,l}) = (2a_{k,l}\sqrt{a_{k,l}^2}) e^{-a_{k,l}^2 \sqrt{a_{k,l}^2}}. \tag{4.15}$$

The phase coefficients are uniformly distributed within the interval $[0, 2\pi)$.

In Fig. 4.3 the principle of the Saleh-Valenzuela model and the exponential decrease of the root mean square value is shown. Based on the S-V model, the IEEE 802.15.3 working group for Wireless Personal Area Networks decided in 2003 to use the so called modified Saleh-Valenzuela model as a reference UWB channel model. In 2006 the IEEE 802.15.4a task group standardized this UWB channel model for simulation and evaluation of UWB systems [95]. In this model equation 4.11 becomes

$$h(t) = \sum_{l=0}^{L} \sum_{k=0}^{K} a_{k,l} e^{j\Phi_{k,l}} \delta(t - T_l - \tau_{k,l}). \tag{4.16}$$

K is the number of MPCs within a cluster and L depicts the number of clusters. L can either be assumed fixed [96] or considered to be a stochastic variable. In [82] the number of clusters is modeled as Poisson distributed with

$$p(L) = \frac{L^L e^{-L}}{L!}. \tag{4.17}$$

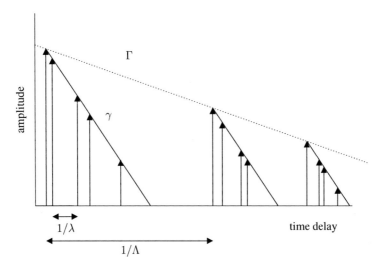

Figure 4.3: Principle of the Saleh-Valenzuela model.

The inter-arrival times of the clusters are still modeled by equation 4.12, where Λ is assumed to be independent of l and $1/\Lambda$ is typically in the range of 10-50 ns [82,84,87,97]. For the interarrival times of the MPCs within a cluster, a mixture of two Poisson processes is used so that equation 4.13 becomes

$$p\left(\tau_{k,l}\middle\|\tau_{k-1,l}\right) = \beta\lambda_1 e^{-\lambda_1\left(\tau_{k,l}-\tau_{k-1,l}\right)} + \left(1-\beta\right)\lambda_2 e^{-\lambda_2\left(\tau_{k,l}-\tau_{k-1,l}\right)}, \qquad (4.18)$$

where β is the mixture probability, and λ_1 and λ_2 are the ray arrival rates.

The next step is the determination of the cluster powers. For scenarios with a LOS condition, the power delay profile of each cluster is given by a one-sided exponential decay

$$E\left\{\left|a_{k,l}\right|^2\right\} \propto \Omega_l e^{-\tau_{k,l}/\gamma_l}, \qquad (4.19)$$

where Ω_l is the integrated energy of the l-th cluster, and γ_l is the intracluster decay time constant. The value for γ differs widely in literature between 1 and 60 ns [82,84,87,98,99] for the standardized channel model, γ_l depends on the cluster arrival time T_l with the linear equation

$$\gamma_l = k_\gamma T_l + \gamma_0, \qquad (4.20)$$

with the constants k_γ and γ_0. The cluster powers follow an exponential decay, if averaged over the large scale fading

$$10\log\left(\Omega_l\right) = 10\log\left(e^{-T_l/\Gamma}\right) + M_{Cluster}, \qquad (4.21)$$

where $M_{Cluster}$ is a random normal distributed variable with standard deviation $\sigma_{Cluster}$. For NLOS conditions the channel model uses only one cluster, whose energy increases at the beginning and decays after a local maximum, which is given by

$$E\left\{|a_{k,1}|^2\right\} \propto \left(1 - \chi e^{-\tau_{k,l}/\gamma_{rise}}\right) e^{-\tau_{k,l}/\gamma_1}. \tag{4.22}$$

Here χ is the attenuation of the first component whereas γ_{rise} depicts the rise before the maximum and γ_1 the following decay.

4.2.4 Small-Scale Fading

Small-scale fading refers to variations in the amplitude of the amplitude coefficients $a_{k,l}$ over a small area caused by the superposition of unresolvable components. In the standardized channel model $a_{k,l}$ is modeled as Nakagami distributed with

$$P\left(|a_{k,l}|\right) = \frac{2}{\Gamma(m)} \left(\frac{m}{\Omega}\right)^m |a_{k,l}|^{2m-1} e^{-\frac{m}{\Omega}|a_{k,l}|^2}, \tag{4.23}$$

where m is the Nakagami m-factor, $\Gamma(m)$ is the Gamma function, and Ω is the root mean square of the amplitude.

4.3 UWB Channel Measurement Campaign in an Industrial Environment

4.3.1 Impulse Response Measurement Techniques

In general channel measurements can be done either in time domain (TD) or frequency domain (FD). The TD approach is based on the impulse transmission and the measuring of the channel impulse response using a digital sampling oscilloscope. The bandwidth depends on the pulse duration and on the oscilloscope's characteristics. The simpler the pulse shape, the easier is the down convolution of the received signal with the transmitted pulse. The corresponding train of impulses can also be generated using conventional direct sequence spread spectrum based measurement systems with a correlation receiver. The drawback of this technique is that to achieve bandwidths of several GHz very high chip rates are needed. The concept of a TD sounder configuration is shown in Fig. 4.4. Channel measurements can also be performed by using a vector network analyzer (VNA). An example setup is shown in Fig. 4.5. In that technique, the sounding signal is a set of narrow-band sinusoids, swept across a wide frequency band and the channel frequency response is recorded with the network analyzer. This corresponds to a measurement of the S_{21}-parameter of a device under test (DUT), which is the radio channel. This FD approach allows the use of broadband antennas, without any special pulse characteristics. The drawback of the FD channel sounding is that due to the long sweep time, the channel has to be nearly static. For fast changing environments, the TD measurement approach should be preferred. When using the VNA method, an upper bound for the maximum detectable delay of the channel in time domain τ_{max} can be defined, which depends on the number of frequency points per sweep N_{sample} and the sweep bandwidth B:

$$\tau_{max} = \frac{N_{sample} - 1}{B} \qquad (4.24)$$

The S-parameters of the channel are measured at each frequency point in the sweep. S_{21} represents the frequency response $H(\omega)$ of the channel. The VNA measures amplitude, phase and via the inverse fast Fourier transform (IFFT) the impulse response (IR) $h(t)$ can be obtained. Theoretically, having a static channel, and occupying an unlimited bandwidth, both techniques end up with the same results.

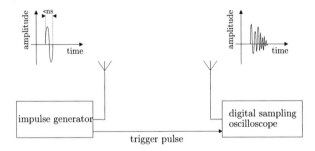

Figure 4.4: Time domain UWB radio channel measurement system.

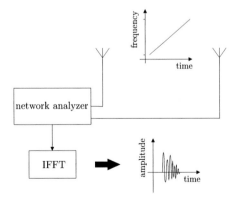

Figure 4.5: Frequency domain UWB radio channel measurement system using a VNA.

4.3.2   Measurement Setup

A channel measurement campaign of UWB channels in industrial environments is done at a typical factory hall. The channel soundings are performed in the frequency domain using a VNA (*Rohde & Schwarz ZVB 8*). The network analyzer is operated in response mode, where PORT 1 is a transmitter and PORT 2 is a receiver port. The sweep time is automatically chosen by the analyzer, depending on the used bandwidth and number of frequency points. The antennas used in this setup are UWB antennas by *Skycross* [100], which have an omni-directional radiation pattern. The antenna gain is specified by the antenna manufacturer. The measured complex S-parameters are stored in the VNA and post processed on a PC using Matlab. Table 4.1 gives the main parameters of the measurement setup. Using these parameters an upper limit τ_{max} for the detectable delay of the channel can be calculated according to Eq. 4.24. This maximum delay corresponds to over 600 m, which is more than needed for an indoor environment. During the measurement process, the physical indoor environment was kept as static as possible, but no effort was made to suppress or control the radio interference of other possible RF sources. The distance between the antennas are varied from 0.5 m to 10 m with line-of-sight (LOS) and none-line-of-sight (NLOS) conditions.

Before measuring, the VNA requires a calibration with the same cables and adapters as will be used in the measurement. The calibration procedure sets the time reference points from the ports of the analyzer, to the end of the used cables. As a result, the obtained power delay profiles include only the propagation delays that are introduced by the radio channel. The calibration also compensates the frequency dependent variations of the transmitted signal level caused by the long cables and adapters.

In addition to the measurement of the channel impulse response, a simple measurement approach is done to obtain a rough estimation of the propagation exponent n of the UWB radio channel in a typical industrial environment. With this parameter the specific PG for that industrial environment can be modeled. The measurement setup used for this estimation is shown in Fig. 4.6. A FMCW transmitter is used to transmit a linear frequency ramp with an output power of 10 dBm. The signal is received by a spectrum analyzer (*Rohde & Schwarz FSUP 26*) and the received signal power is observed. A reference measurement is performed at a distance d of 0.5 m between the transmitting and the receiving antenna. The distance d is successively increased and with the obtained received signal power, the propagation exponent can be approximated.

Parameter	Value
Frequency band	6 GHz - 8 GHz
Bandwidth	2 GHz
Number of frequency points	4096
Sweep time	4.2 s
Average noise floor	-100 dBm
Transmitted power	0 dBm

Table 4.1: VNA measurement setup parameters.

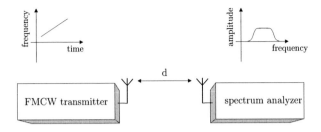

Figure 4.6: Measurement setup for estimating the path gain.

4.3.3 Measurement Results and Conclusion

Path Gain

The results of the measurements of the path gain are shown in Fig. 4.7. 15 measurements were made with different distances between the transmitting and the receiving antenna. Although this approach is not that accurate, a rough estimation of the specific propagation exponent and therefore the according PG of that environment is obtained. To estimate the propagation exponent of that specific environment, a curve fitting algorithm is applied to the measured values. This curve fitting is done using the linear least squares method. The fitted curve is also shown in Fig. 4.7 and results in a line with a gradient of -1.922 when plotted on a logarithmic scale which denotes a propagation exponent of 1.922. Compared to the propagation exponent values given in section 4.2.1, this corresponds more to the model for office and residential environments than typical industrial environments. This can be explained by the very large proportions of the production hall with no narrow corridors, so that the propagation exponent is close to the free space exponent of 2.

Channel Impulse Response

As mentioned before, the channel sounding is performed with a number of different antenna positions, which could be seen as typical scenarios for a local positioning system for flexible tools. The antenna constellations for these measurements are plotted in the schematic floor plan in Fig. 4.8. The first measurement is made with the transmitting antenna located inside the autobody (a) with a LOS through the side window. The second position is located near the rear wheelhouse (b), the antenna positions (c) and (d) are again located inside the car body at the driver's and the co-driver's seat. Antenna position (e) is aside of the assembly line and position (f) is located at the front of the autobody, near the engine bay. The impulse responses for these six different measurement conditions are shown in Fig. 4.9.

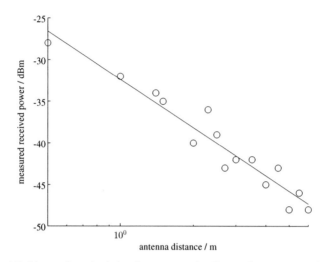

Figure 4.7: Measured received signal power over the distance between transmit and receive antenna and linear fitted curve. The resulting propagation exponent is close to the free space exponent of 2.

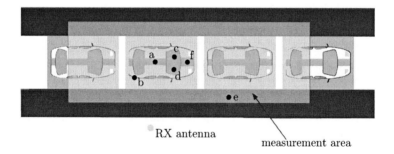

Figure 4.8: Antenna constellations for six exemplary measurement scenarios. The receiving antenna is located at the side of the assembly line. The transmit antenna is placed at various positions (marked with characters a-e).

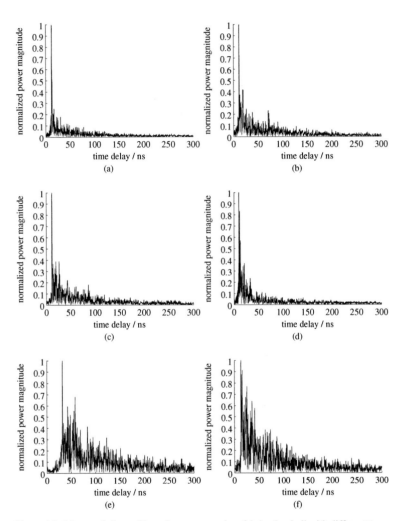

Figure 4.9: Measured channel impulse responses in a fabrication hall with different transmitting antenna positions. (a) Transmitting antenna inside autobody, (b) near autobody wheelhouse, (c) inside autobody at driver's seat, (d) inside autobody at co-driver's seat, (e) along the assembly line, and (f) inside autobody engine bay.

The figures show that the channel impulse responses in this environment are extremely dependent on the specific antenna constellation. In general it can be stated that all measured channel impulse responses contain a significant amount of multipath components close to the LOS signal. The absolute number of echoes an their respective signal power vary widely. In contrast to the standardized UWB channel model for industrial environments presented in section 4.2, a generalization of that environment is not possible. Hence, the channel model has to be adapted for every measurement constellation under investigation by simulation.

Since the first echoes close to the LOS signal are the most critical ones for the performance of an indoor positioning system, the mean time delay between these two signals for typical tool positions are investigated. The calculated mean delay of 40 measurements is 1.7 ns. For an indoor FMCW based positioning system, this time delay has to be resolvable to work sufficiently in that multipath environment.

CHAPTER 5

Pulsed Frequency Modulated Ultra-Wideband Positioning System

In the following chapter, the developed Pulsed Frequency Modulated Ultra-Wideband Positioning System is introduced. The basic principles were first presented in [101–104]. The system uses a combination of common FMCW radar technique and a pulsed UWB approach. The first section gives a short overview of the LPR system developed by the *Symeo GmbH*, which is the basis of the later UWB system. This section is followed by an investigation of the limitations of the LPR system due to multipath signal propagation and how this problem can be solved. Subsequently the expansion of the narrowband system to an UWB system by extending the sweep bandwidth and introducing a fast RF switch to chop the transmit signal is shown. This is followed by an analysis of the complete receiver chain. The chapter concludes with a detailed description of the implemented hardware demonstrator.

5.1 The Symeo Local Positioning Radar System

The Pulsed Frequency Modulated UWB Positioning System is based on the LPR system developed by the *Symeo GmbH* in Munich, Germany [105]. In this system, the round-trip time-of-flight of a signal traveling from one station to another and back is measured and the distance between these units is calculated. To maximize the operating distance, the conventional frequency-modulated continuous-wave radar approach was adapted to a system of two active units, as depicted in Fig. 5.1. In the following a brief introduction to the functional principle of the *Symeo* LPR is given. For a detailed discussion it is referred to [60, 106–110].

The distance measurement with the LPR system is performed in two steps. First, the mobile client has to be synchronized to the base station with high precision and afterwards the distance between the two stations can be measured.

Figure 5.1: Setup of the Symeo secondary radar system according to [60]. The first unit (base station: BS) transmits a signal to the second unit (mobile client: MC). The mobile client synchronizes to the impinging signal and sends a synchronized reply back to the base station.

Synchronization Principle

For the synchronization, as well as for the distance measurement, linear frequency modulated signals are used. These signals are specified by the start frequency of the sweep f_l, the stop frequency f_u, the sweep bandwidth $B = f_u - f_l$, the center frequency $f_c = f_l + \frac{B}{2}$, the sweep time T, and the sweep rate μ which is the ratio B/T. In order to synchronize the modules, the base station transmits the FMCW signal starting at time $t = 0$. This signal $x_{U1,tx}$ is represented by

$$x_{U1,tx} = A_{U1,tx} \cos\left(2\pi f_c t + \pi\mu t^2 - \pi\mu T t + \varphi_0\right), \tag{5.1}$$

where $A_{U1,tx}$ is the amplitude of the signal and φ_0 is an arbitrary phase term. After a delay of t_d due to the time-of-flight the signal arrives at the mobile client. The received signal $x_{U2,rx}$ is a delayed and attenuated copy of the transmitted signal and is given by

$$x_{U2,rx} = A_{U2,rx} \cos\left(2\pi f_c \left(t - t_d\right) + \pi\mu \left(t - t_d\right)^2 - \pi\mu T \left(t - t_d\right) + \varphi_1\right)$$
$$\forall\, t \in \left(t_d + \Delta t, t_d + T\right), \tag{5.2}$$

where $A_{U2,rx}$ is the amplitude of the received signal and φ_1 an arbitrary constant phase term. In the mobile client a local signal is generated which can be expressed by

$$x_{U2,lo} = A_{U2,lo} \cos\left(2\pi \left(f_c + \Delta f\right)\left(t - t_d - \Delta t\right) + \pi\mu \left(t - t_d - \Delta t\right)^2\right.$$
$$\left. - \pi\mu T \left(t - t_d - \Delta t\right) + \varphi_2\right)$$
$$\forall\, t \in \left(t_d + \Delta t, t_d + \Delta t + T\right). \tag{5.3}$$

$A_{U2,lo}$ is the amplitude of the local signal of the mobile client and φ_2 another arbitrary constant phase term respectively. Fig. 5.2 shows the respective frequency of the received signal $x_{U2,rx}$ and the locally generated signal $x_{U2,lo}$ of the mobile client. The local signal has a slightly different starting point in time and frequency and has to be synchronized to $x_{U2,rx}$ with respect to the offsets Δt and Δf. These offsets can be estimated by multiplying the received and the locally generated signals in a mixer. In accordance to Eq. 5.2 and 5.3 the result can be calculated with

$$x_{U2,mix} = x_{U2,lo} x_{U2,rx}$$
$$= \frac{A_{U2,lo} A_{U2,rx}}{2} \left(\cos\left(\varphi_{U2,lo} + \varphi_{U2,rx}\right) + \cos\left(\varphi_{U2,lo} - \varphi_{U2,rx}\right)\right) \tag{5.4}$$

with

$$\varphi_{U2,lo} = 2\pi \left(f_c + \Delta f\right) \left(t - t_d - \Delta t\right) + \pi\mu \left(t - t_d - \Delta t\right)^2 - \pi\mu T \left(t - t_d - \Delta t\right) + \varphi_2 \tag{5.5}$$

and

$$\varphi_{U2,rx} = 2\pi f_c \left(t - t_d\right) + \pi\mu \left(t - t_d\right)^2 - \pi\mu T \left(t - t_d\right) + \varphi_1 \tag{5.6}$$

respectively.

This mixing product is low-pass filtered resulting in an intermediate frequency (IF) signal with a constant frequency, which can be expressed by

$$x_{if} = A_{if} \cos\left(\varphi_{U2,lo} - \varphi_{U2,rx}\right), \tag{5.7}$$

where A_{if} is a constant depending on the amplitudes of the mixed signals. The frequency of the signal x_{if} is obtained by differentiating its phase with respect to time t:

$$f_{if} = \frac{1}{2\pi}\frac{d}{dt}\left(\varphi_{U2,lo} - \varphi_{U2,rx}\right) \tag{5.8}$$

which calculates to

$$f_{if} = \Delta f - \frac{B}{T}\Delta t \qquad\qquad \forall t \in \left(t_d + \Delta t, t_d + T\right). \tag{5.9}$$

By estimating this frequency during the up- and downsweep, two frequencies f_1 and f_2 are obtained. With these two frequencies the linear set of equations

$$f_1 = \Delta f - \frac{B}{T}\Delta t \qquad\qquad \forall t \in \left(t_d + \Delta t, t_d + T\right) \tag{5.10}$$

$$f_2 = \Delta f + \frac{B}{T}\Delta t \qquad\qquad \forall t \in \left(t_d + \Delta t + T, t_d + 2T\right) \tag{5.11}$$

can be solved with respect to Δf and Δt

$$\Delta f = \frac{f_2 + f_1}{2} \tag{5.12}$$

$$\Delta t = \frac{T}{B}\frac{f_2 - f_1}{2}, \tag{5.13}$$

and the offsets can be corrected. In practice, the frequencies f_1 and f_2 are obtained by sampling the low-pass filtered mixing product during the up- and downsweep of the locally generated frequency ramps, applying an fast Fourier transform (FFT) algorithm and evaluating the obtained signal spectrum.

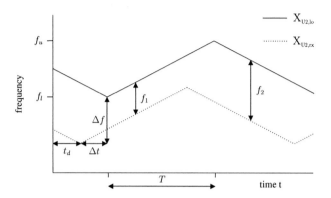

Figure 5.2: Frequency of the locally generated ($x_{U2,lo}$) and the received signal ($x_{U2,rx}$) of the mobile client during the synchronization. The two signals are multiplied and after low-pass filtering, the frequencies f_1 and f_2 are linearly dependent on the offsets Δf and Δt.

Distance Measurement

After a successful synchronization, the local generated signal $x_{U2,lo}$ of the mobile client matches the received signal $x_{U2,rx}$ exactly in time and frequency. The mobile client sends the synchronized signal with a known offset in time and frequency back to the base station. This signal arrives at the first module with a delay corresponding to the round-trip time-of-flight $2t_d$. For the measurement of the distance between the two stations, a single up- or downsweep is sufficient. Fig. 5.3 shows the respective frequency of the locally generated frequency ramp $x_{U1,lo}$ of the base station during the downsweep and the frequency of the received signal $x_{U1,rx}$. The locally generated and the received signal are multiplied in a mixer followed by a low-pass filter. The frequency of the filtered mixing product for an up- and a downsweep are given by

$$f_{if,up} = \Delta f_{a,up} - 2t_d \frac{B}{T} \tag{5.14}$$

$$f_{if,dwn} = \Delta f_{a,dwn} + 2t_d \frac{B}{T} \tag{5.15}$$

and are linearly dependent on the round-trip time-of-flight. The distance d between the two modules can be calculated from either $f_{if,up}$ or $f_{if,dwn}$ with

$$d = \frac{c_0 T}{2B} \left(\Delta f_{a,up} - f_{if,up} \right) \tag{5.16}$$

$$d = \frac{c_0 T}{2B} \left(f_{if,dwn} - \Delta f_{a,dwn} \right). \tag{5.17}$$

The frequency offsets $\Delta f_{a,up}$ and $\Delta f_{a,dwn}$ are known system parameters and are introduced to ensure a positive frequency of the IF-signal.

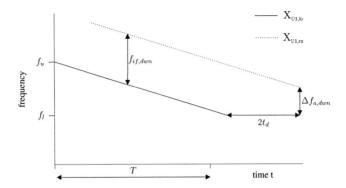

Figure 5.3: Frequency of the locally generated ($x_{U1,lo}$) and the received signal ($x_{U1,rx}$) of the base station during the distance measurement at the downsweep. The two signals are multiplied and after low-pass filtering, the frequency $f_{if,dwn}$ is linearly dependent on the round-trip time-of-flight $2t_d$ and therefore on the distance between the two units d.

The system parameters for the *Symeo* LPR are given in table 5.1. The system allows a very precise distance measurement with a standard deviation of the measured distances as low as 1 cm, implying a synchronization of both units in time and frequency of less than 100 ps and 10 Hz, respectively. By adding multiple base stations it is possible to build up a three-dimensional wireless local positioning system with an 3D-accuracy in the range of 15 cm. Furthermore the Doppler frequency shift of the radar signals is used to measure the relative velocity of the radar stations [110].

System Parameter	Value
Sweep Bandwidth B	150 MHz
Sweep Time T	1 ms
Center Frequency	5.8 GHz
Max. Operation Range	2.2 km
1-D Standard Deviation	< 1 cm

Table 5.1: *Symeo* LPR system parameters.

5.2   Multipath Issues

Propagation over multiple signal paths is a well known problem in microwave transmissions. The *Symeo* LPR system performs excellent in a free-space or moderate indoor channel but especially in dense multipath environments, the system performance of the measurement system is limited by multipath distortions. Assuming a successful synchronization of both units, the base station receives a superposition of multiple time-delayed copies of the response signal which correspond to the different multipaths. Each multipath component has its individual distance and path loss, resulting in different amplitudes and time delays. Due to the FMCW radar principle, each time delay is transferred to a certain frequency. The low-pass filtered baseband signal of Eq. 5.7 gets more complicated and can be expressed as an additive superposition of multiple cosine functions:

$$x_{if}(t) = A_{LOS} \cos\left(2\pi f_{LOS} t + \Phi_{LOS}\right) + \sum_i A_{NLOS,i} \cos\left(2\pi f_{NLOS,i} t + \Phi_{NLOS,i}\right),$$

$$\forall 0 \le t \le T, \tag{5.18}$$

with

$$f_{LOS} < f_{NLOS,i}, \quad \forall i \tag{5.19}$$

during the downsweep and

$$A_{LOS} > A_{NLOS,i}, \quad \forall i \tag{5.20}$$

due to the higher path loss caused by the longer distance of the multipath components compared to the LOS. In Fig. 5.4 the principal mechanism is illustrated. The frequency of the LOS signal is chosen to be $f_{LOS} = 100\,\text{kHz}$, the sweep bandwidth B is given with $150\,\text{MHz}$ according to the LPR parameters. The time delay between the LOS signal and the two multipaths is $7.4\,\text{ns}$ and $20\,\text{ns}$, which results in the two additional frequency components with $f_{NLOS,1} = 102.22\,\text{kHz}$ and $f_{NLOS,2} = 106\,\text{kHz}$ respectively. For the sake of simplicity, the amplitudes of the signal echoes are chosen to be $3/4$ and $1/2$ of the amplitude of the LOS signal. It can be seen, that these additional signals in the frequency spectrum lead to a distortion of the first peak. The first two maxima are shifted towards each other and do not correspond to the correct frequency values anymore. This error in the frequency estimation of the LOS signal translates directly into an error during the distance calculation.

The distortions are analyzed by a simple example using two complex signals

$$x_{LOS}(t) = e^{j2\pi f_{LOS} t} e^{j\Phi_{LOS}} w\left(t - \frac{T}{2}\right), \qquad \forall 0 \le t \le T \tag{5.21}$$

$$x_{NLOS}(t) = e^{j2\pi f_{NLOS} t} e^{j\Phi_{NLOS}} w\left(t - \frac{T}{2}\right), \qquad \forall 0 \le t \le T \tag{5.22}$$

The amplitudes for both signals are set to 1. The parameter w represents the window function which ensures a time limit of the signals and is assumed to be even and real-valued. It is shifted by $T/2$ to the right. The two signals correspond to the direct LOS and one multipath NLOS component. The Fourier transform of these signals can be calculated with

$$X_{LOS}(f) = e^{j\Phi_{LOS}} W\left(f - f_{LOS}\right) e^{-j2\pi(f - f_{LOS})\frac{T}{2}} \tag{5.23}$$

$$X_{NLOS}(f) = e^{j\Phi_{NLOS}} W\left(f - f_{NLOS}\right) e^{-j2\pi(f - f_{NLOS})\frac{T}{2}} \tag{5.24}$$

with W as the Fourier transform of the window function w.

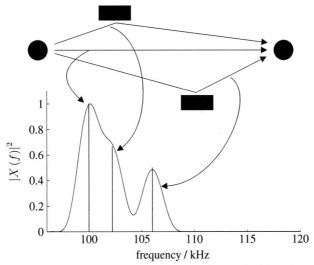

Figure 5.4: Illustration of multipath propagations. The signal emitted from the transmitter reaches the receiver on multiple paths with increasing delays. The individual paths give rise to additional peaks in the IF signal spectrum of the FMCW radar system. The first maximum of the spectrum does not correspond to the frequency of the LOS signal and an error in the distance calculation is introduced.

Because of the assumptions made for the window function w in time domain, the Fourier transform of the window function W is also even and real-valued [111]. Because of the linearity of the Fourier transformation the sum of the two signals $x(t) = x_{LOS}(t) + x_{NLOS}(t)$ is transformed to the sum of $X(f) = X_{LOS}(f) + X_{NLOS}(f)$. The energy density spectrum can then be calculated with

$$\begin{aligned}
|X(f)|^2 &= X(f)X(f)^* \\
&= \left(e^{j\Phi_{LOS}} W(f - f_{LOS}) e^{-j2\pi(f-f_{LOS})\frac{T}{2}} + e^{j\Phi_{NLOS}} W(f - f_{NLOS}) e^{-j2\pi(f-f_{NLOS})\frac{T}{2}} \right) \\
&\quad \cdot \left(e^{-j\Phi_{LOS}} W(f - f_{LOS}) e^{j2\pi(f-f_{LOS})\frac{T}{2}} + e^{-j\Phi_{NLOS}} W(f - f_{NLOS}) e^{j2\pi(f-f_{NLOS})\frac{T}{2}} \right) \\
&= \left(W(f - f_{LOS})^2 + W(f - f_{NLOS})^2 + W(f - f_{LOS}) W(f - f_{NLOS}) \right) \\
&\quad \cdot \left(e^{j(\Phi_{LOS}) - \Phi_{NLOS}} e^{j2\pi(f_{LOS}-f_{NLOS})\frac{T}{2}} + e^{-j(\Phi_{LOS}-\Phi_{NLOS})} e^{-j2\pi(f_{LOS}-f_{NLOS})\frac{T}{2}} \right) \\
&= \left(W(f - f_{LOS})^2 + W(f - f_{NLOS})^2 + W(f - f_{LOS}) W(f - f_{NLOS}) \right) \\
&\quad \cdot 2\cos\left(\Phi_{LOS} - \Phi_{NLOS} + 2\pi(f_{LOS} - f_{NLOS})\frac{T}{2} \right).
\end{aligned} \tag{5.25}$$

Since the cosine function is an even function the following equation applies:

$$|X(f)|^2 = W(f - f_{LOS})^2 + W(f - f_{NLOS})^2 + W(f - f_{LOS})W(f - f_{NLOS})$$
$$\cdot 2\cos\left(\Phi_{NLOS} - \Phi_{LOS} + 2\pi(f_{NLOS} - f_{LOS})\frac{T}{2}\right). \quad (5.26)$$

The complete energy density spectrum consists of three additive parts. The first two represent the energy density spectra of the window function shifted to f_{LOS} and f_{NLOS}. The third part represents the product of the shifted spectra weighted by a cosine function. Due to the additive superposition of these three parts, the energy density spectrum is distorted. To reduce the total distortion, all three parts have to be minimized. The two requirements

$$|X(f_{LOS})|^2 \overset{!}{=} W(0)^2 \quad (5.27)$$
$$|X(f_{NLOS})|^2 \overset{!}{=} W(0)^2 \quad (5.28)$$

lead to the necessary and sufficient condition for a minimum distortion of the energy density spectrum:

$$W(f_{NLOS} - f_{LOS})^2 = W(f_{LOS} - f_{NLOS})^2 \approx 0. \quad (5.29)$$

The bandwidth of the window function is determined by the duration of the time signal and the kind of window function. In a given system, these parameters are constant. The difference $f_{NLOS} - f_{LOS}$ corresponds with the difference of the two signal paths $d_{NLOS} - d_{LOS}$. Therefore a minimum allowable path difference can be calculated, where equation 5.29 is valid and the maxima of the energy density spectrum are not distorted:

$$\Delta d_{min} = d_{NLOS} - d_{LOS}. \quad (5.30)$$

According to equations 5.16 and 5.17 this minimum distance is inversely proportional to the applied sweep bandwidth of the system B. The ability to resolve closely spaced paths therefore depends on three parameters:

• The type of window function,

• The FFT bin size (frequency domain resolution) and

• The system sweep bandwidth B.

While the first two parameters are limited by a Δd_{min}, the increase of the sweep bandwidth leads to an increase of the differences between the frequency components, making them resolvable. In Fig. 5.5, the effect of increasing B is illustrated with four different sweep bandwidths. All other system parameters were chosen according to the system settings of the *Symeo* LPR narrowband system. It is assumed that the signal of the LOS path is mapped to an IF signal with a frequency of 100 kHz. The NLOS signals arrive at the receiver over longer distances with $\Delta d_1 = 0.9$ m, $\Delta d_2 = 2.1$ m, and $\Delta d_3 = 3.0$ m and cause additional peaks in the IF spectrum. The figures show the expected peak positions as vertical lines. It can be seen that in the $B = 150$ MHz case, only two peaks are clearly resolved. In addition to that, the first peak is shifted to the left so that its maximum does not correspond to the frequency of the LOS signal. This has an immediate impact on the total accuracy of the localization system. In this example the frequency error when determining the first maximum as the LOS peak is 176.6 Hz, which leads to an error of 17 cm in the calculation of the distance. By increasing the sweep bandwidth, the multipath components are mapped to higher frequencies, increasing the difference to the LOS peak and therefore the resolution capability. With a bandwidth of $B = 1$ GHz all four peaks are clearly resolved and the remaining difference between the maximum of the first peak and the real LOS peak decreases to 1 Hz.

Since the correct detection of the direct path signal in the presence of dense multipath channel characteristic determines the accuracy of a ranging and positioning system, it is therefore obvious that for an enhanced multipath robustness a large sweep bandwidth should be chosen. Such large signal bandwidths are only available for UWB systems which allow a very fine time resolution and therefore resolution of multipath components. In general it can be stated the higher the bandwidth, the better the robustness towards distortions introduced by mutlipath signal components.

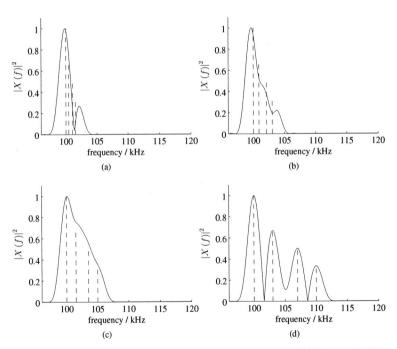

Figure 5.5: The effect of separability (resolution) of paths, exemplified on a simplified multipath scenario with three multipath components. (a) $B = 150$ MHz (b) $B = 300$ MHz (c) $B = 500$ MHz (d) $B = 1$ GHz.

5.3   Expansion to Ultra-Wideband FMCW

The primary goal of the ultra-wideband positioning system for complex environments (UPos) research project was to develop an UWB system, based on the well known LPR architecture, and to get this system in accordance with the international emission regulations. Since it has been shown in the previous section that in a common secondary radar based positioning system, the increase of the sweep bandwidth is an obvious way to enhance the multipath resolution, investigations with different frequency setups have been made. The results of the channel measurement campaign presented in chapter 4 lead to the conclusion that a minimum sweep bandwidth of 1 GHz should be chosen to reach acceptable position estimations in an industrial environment. To fit in the US FCC frequency mask, as well as in the European ECC frequency mask, a system to be developed has to work between 6 GHz and 8.5 GHz, since this is the only common frequency range with the maximum allowed transmit power. Fig. 5.6 shows the simulation result of the average power spectrum density of a frequency sweep from 7 GHz to 8 GHz with a continuous wave (CW) signal power of 0 dBm and the appropriate settings for the measurement of the average power spectral density of UWB signals as described in appendix A. Since an resolution bandwidth (RBW) of 1 MHz and an averaging time of 1 ms has to be applied, a frequency sweep of 1 GHz and a sweep time of 1 ms results in an averaging factor of $1/1000$ which corresponds to an average power spectral density of the FMCW signal as low as -30 dBm/MHz. Also shown in this figure is the emission limit given by the FCC. It can be seen that the average power density of the radar signal exceeds the -41.3 dBm/MHz limit by 11.3 dB.

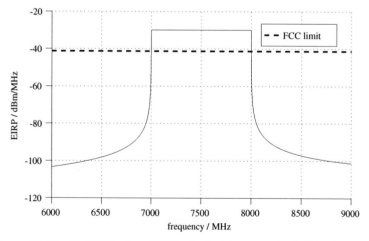

Figure 5.6: Simulated EIRP of an FMCW signal with a sweep bandwidth of 1 GHz, a sweep time of 1 ms, and a peak power of 0 dBm. The average PSD is calculated according to the UWB measurement specifications and violates the emission limit by 11.3 dB.

Furthermore a simple sweep over a large bandwidth ($> 500\,\text{MHz}$) does not qualify a system as an UWB system according to the FCC rules, as described in chapter 2. Hence, the FMCW signal has to be modified in a way that the average power density drops below the given emission limits and in addition to that, the instantaneously occupied spectrum of the signal has to be expanded over at least $500\,\text{MHz}$. This is done by chopping the FMCW signal into small sections, which results in a multiplication of the FMCW signal with a pulse train. There are different techniques to combine pulses with the FMCW signal. One possibility is that pulses modulate the signal through a mixer [16, 112]. Another, more easier way, is to insert a dedicated fast RF switch, which is driven by a pulse generator. The basic concept of the usage of an RF switch as chopper is shown in Fig. 5.7.

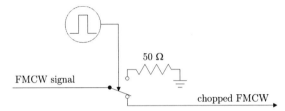

Figure 5.7: Principle of chopping the FMCW signal. A dedicated fast RF switch, which is driven by a pulse generator switches the signal on and off, resulting in a multiplication of the FMCW signal with a pulse train.

Due to the finite rise and fall times of the RF switch, the chopping of the frequency modulated signal can be approximated by a multiplication of the FMCW signal with a Tukey window shaped pulse train. The pulse can be expressed by

$$w\left(t\right) = \begin{cases} 1.0 & ,0 \leq |t - t_0| \leq \alpha\frac{T_w}{2} \\ \frac{1}{2}\left(1 + \cos\left(\pi\frac{(t-t_0)-\alpha T_w/2}{2(1-\alpha)T_w/2}\right)\right) & ,\alpha\frac{T_w}{2} \leq |n| \leq \frac{T_w}{2}. \end{cases} \quad (5.31)$$

The real switching pulse can be modeled correctly by the appropriate choice of α which determines the rise and fall time of the pulse. Fig. 5.8 shows a switching pulse measured with a fast real-time oscilloscope together with a modeled switching pulse according to Eq. 5.31. A good agreement between measurement and the Tukey model can be observed. To obtain the pulse train, the single pulse $w\left(t\right)$ is convolved with a Dirac comb with a periodic time of T_{pp} given with

$$\Delta\left(t\right) = \sum_{k=0}^{\infty} \delta\left(t - kT_{pp}\right). \quad (5.32)$$

Hence, the pulse train for the switching pulses can be expressed by

$$p\left(t\right) = w\left(t\right) * \Delta\left(t\right). \quad (5.33)$$

The result of the multiplication of the Tukey shaped pulse train with the FMCW signal in time domain is depicted in Fig. 5.9.

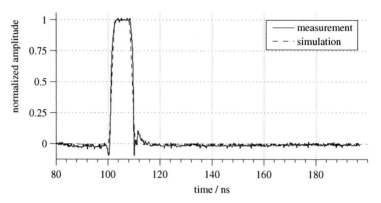

Figure 5.8: Measured switching pulse compared to the used Tukey window model.

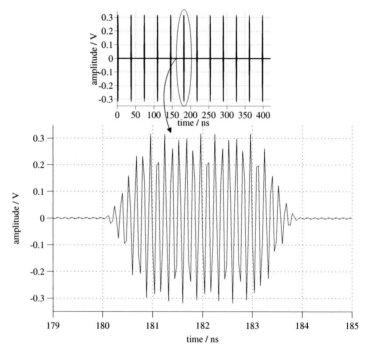

Figure 5.9: Simulated chopped FMCW time domain signal, using an RF switch with a rise and fall-time of 1 ns and an on-time of 2 ns.

The interval between the 50% points of the final amplitude is used to determine the pulse width t_{pw}. The total pulse duration can be calculated via $t_{\text{total pulse}} = t_{rise} + t_{on} + t_{fall}$. The ratio of pulse width to pulse period T_{pp} defines the drop in the average power density of the linearly modulated signal. This drop in decibel is given by

$$P_{drop} = 20 \log_{10} \left(\frac{t_{pw}}{T_{pp}} \right), \qquad (5.34)$$

for an ideal switch with a rectangular switching characteristic and no insertion loss. To decrease the average power density by -11.3 dB, a ratio of $1/3.7$ is necessary. However, due to the finite rise and fall times as well as the finite isolation of the switch, the drop is a few dB smaller than the theoretical value. Simulations with realistic models of RF switches show that to achieve a decrease of -11.3 dB, a ratio of approximately $1/10$ has to be used. In addition to the decrease of the average signal power, the signal is spread over a large frequency range. The spreading factor can be approximated by the reciprocal of the pulse on-time $1/t_{on}$. Fig. 5.10 shows the simulation output of a FCC conform system. The switch is modeled via the presented Tukey window method with a rise and fall time of 1 ns and a pulse width of $t_{pw} = 3$ ns. The resulting instantaneously occupied 10 dB bandwidth is larger than 500 MHz and qualifies the system as an UWB system as defined by the FCC. The ratio t_{pw}/T_{pp} for this simulation was chosen to be $1/12$ which results in a drop of the average power density that is sufficient for getting in compliance with the given emission mask.

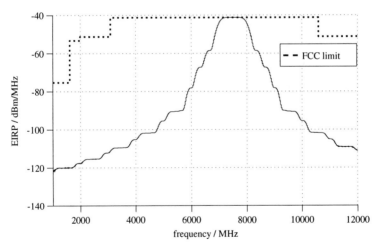

Figure 5.10: Simulation result of the EIRP of a linear frequency modulated signal with a sweep bandwidth of 1 GHz, a sweep time of 1 ms, and a peak power of 0 dBm. The signal is chopped via a fast RF switch with a pulse width of $t_{pw} = 3$ ns and a pulse width to pulse period ratio of $t_{pw}/T_{pp} = 1/12$.

5.4   Signal Reconstruction

In the receive path the chopped FMCW signal has to be reconstructed. This is achieved by multiplying the received chopped RF signal with a locally generated continuous signal. This is done by feeding the received signal to a mixer which is driven by the locally generated continuous RF signal. This results in an intermediate frequency mixing product that is still chopped. It is then filtered with a low-pass filter with a significantly smaller bandwidth than the pulse repetition frequency. The filter is implemented by an active 5th order LC filter shown in Fig. 5.11. This low-pass filtering eliminates all higher frequency mixing products including the rectangular signal parts that have been generated by the RF switch. The result is a continuous sinusoidal intermediate frequency signal without the chopping characteristic, which can easily be digitized. In Fig. 5.12 the intermediate frequency signal in time domain is shown, before and after the low-pass filtering. It can be seen that the filtering extracts the non-chopped fundamental signal out of the mixing product, but also leads to a smaller amplitude of the baseband signal. This drawback can partly be compensated by using an active filter architecture. However, the overall receiver sensitivity is still limited by this technique. As a consequence, the maximum operation range of the wireless pulsed frequency modulated UWB ranging and positioning system decreases. Since the low-pass filter performs an averaging on the chopped IF signal, the reduced average power has to be used for calculating the theoretical limit for the maximum operating range, instead of using the actual transmit peak power. This means for an FCC compliant PFM-UWB system -11.3 dBm instead of 0 dBm.

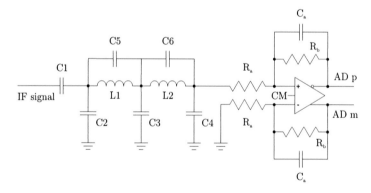

Figure 5.11: Low-pass filter in the receiver chain after the downconverter. This active filter eliminates the higher frequency mixing products as well as the chopping in the RF signal. The result is an differential IF signal in the range of several kHz, which is easy to digitize.

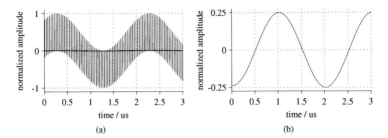

Figure 5.12: Downconverted IF signal in time domain. Before low-pass filtering (a). The on and off state of the signal is clearly visible, after low-pass filtering without amplification (b). The resulting IF signal is a continuous sinusoidal signal in the range of several kHz. This simple method has the drawback of an high loss in the signal amplitude due to the averaging of the signal done by the filter.

5.5 PFM-UWB System Design

In the following, the design and implementation of the PFM-UWB system is described. In Fig. 5.13 a generalized overview of the complete UWB radar module is depicted. The base stations, as well as the mobile units feature an identical hardware setup and can be programmed to work either in base station or in mobile client mode by software. In general the system can be subdivided into four main parts:

- System control

- Radar signal synthesizer

- Receiver chain

- UWB antenna

The overall system control is performed by a digital signal processor (DSP). The processor switches the system between transmit mode, where the generated RF signal is fed to the antenna, and receive mode, where the received signal of the antenna is fed to the mixer as well as the local generated signal. The DSP is also used for the digital signal processing and the calculations during synchronization and distance measurement and must therefore allow for a precise control of the radar signal synthesizer and the timing of the overall measurement process.

The radar signal synthesizer generates the linear frequency modulated radar signals. It is also controlled by the DSP, which programs the start and stop frequency, as well as the sweep bandwidth and sweep time of the frequency ramps and corrects the offsets in time and frequency during the synchronization process. Its performance is most critical for the overall system performance.

In the receiver chain, the received signal is amplified, filtered, downconverted and digitized. Since the low transmit power of UWB signals, a high sensitive receiver architecture is necessary to allow a sufficient system operation range.

As interface to the air, the antenna is a critical element in the signal flow of the UWB radar system. Because of this, various antenna structures are examined and tested for the use in the PFM-UWB system.

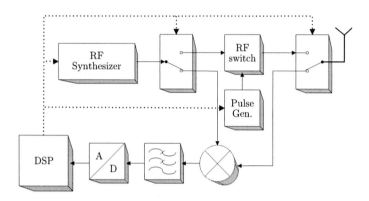

Figure 5.13: Block diagram of the complete radar module. It consists of a DSP, the radar signal synthesizer for providing the chopped FMCW signal, the receiver chain and an UWB antenna. In transmit mode, the output of the RF synthesizer is fed to an fast RF switch, chopping the FMCW signal before it is fed to the antenna. In receive mode, the received signal is fed to a mixer, driven by the non-chopped locally generated RF signal. The mixing product is then low-pass filtered, digitized and further processed by the DSP.

5.5.1   Radar Signal Synthesizer

The signal generator for the pulsed frequency modulated ultra-wideband system consists of two main parts, the FMCW generator and the chopper. Fig. 5.14 shows a schema of the implemented signal synthesizer.

The Phase Locked Loop Synthesizer

The FMCW frequency ramps are generated with a phase-locked loop (PLL) synthesizer. A phase frequency detector (PFD) compares a reference signal to the output signal of a voltage controlled oscillator (VCO) after this output signal is fed through a frequency divider with a divider ratio of $1/N_{pll}$. The PFD produces an error signal, which is proportional to the phase difference of the two input signals. For the PFM-UWB system, the reference signal is generated by a direct digital synthesizer (DDS) and divided by the PLL reference divider R_{ref}.

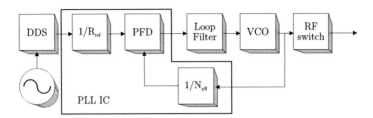

Figure 5.14: Block diagram of the signal generator. The DDS generates the reference signal for the PLL, which consists of a dedicated PLL IC, an active loop filter and a VCO. The synthesizer is programmed to generate frequency sweeps between 7 GHz and 8 GHz. The signal output of the VCO is fed to a fast RF switch, which chops the RF signal and generates a regulation conform UWB signal.

The error signal is filtered by an active low-pass filter and drives the tuning input of the VCO in order to reduce the error. The PLL is locked if the phases of the reference signal and the divided VCO signal are equal. The reference signal is derived from a DDS. In the locking state the output frequency of the VCO is proportional to the reference frequency of the DDS according to

$$f_{VCO} = f_{DDS} \frac{N_{pll}}{R_{ref}}.$$ (5.35)

The DDS is programmed to generate frequency up and down ramps from 35 MHz to 40 MHz. The frequency dividers are programmed to be $N_{pll} = 200$ and $R_{ref} = 1$ so that the resulting frequency ramps are located in the middle of the approved UWB frequency spectrum from 7 GHz to 8 GHz.

One important parameter of the signal synthesizer is the phase noise of the generated RF signal. Phase noise is the frequency domain representation of fast, short-term, random fluctuations in the phase of the RF signal, caused by time domain instabilities. According to [113] the total phase noise $L_{total}(f)$ at the output of the PLL can be described by

$$L_{total}(f) = 10 \log \left(10^{\frac{L_{PLL}(f)}{10}} + 10^{\frac{L_{VCO}(f)}{10}} \right.$$
$$\left. + 10^{\frac{L_{TCXO}(f)}{10}} + 10^{\frac{L_{Resistor}(f)}{10}} \right).$$ (5.36)

The term $L_{PLL}(f)$ represents the influence of the dedicated PLL-IC, described in [113] in detail. $L_{VCO}(f)$ gives the total phase noise contribution of the used VCO. The values can be taken from the VCO's data sheet, where normally the phase noise is approximated by a linear decrease over the frequency in a double logarithmical diagram. The corresponding transfer function has a high-pass characteristic and is given by

$$L_{VCO}(f) = L_{\text{VCO @ 10kHz}} - 30 \log\left(\frac{f}{10kHz}\right)$$
$$+ 20 \log\left|\frac{1}{1 + G(j2\pi f) H}\right|, \tag{5.37}$$

with $G(j2\pi f)$ as the loop filter transfer function and H as the PLL feedback, $1/N_{pll}$. $L_{\text{VCO @ 10kHz}}$ gives the phase noise of the VCO at an offset of 10 kHz.

$L_{TCXO}(f)$ is the noise contribution of the reference, in this case generated by a DDS. The noise characteristic is again taken from the DDS data sheet and weighted with a transfer function with a low-pass characteristic:

$$L_{TCXO}(f) = L_{\text{TCXO @ 10kHz}} - 10 \log\left(\frac{f}{10kHz}\right)$$
$$- 20 \log\left(\frac{f_{Ref}}{f_{Comp}}\right) + 20 \log\left|\frac{G(j2\pi f)}{1 + G(j2\pi f) H}\right|. \tag{5.38}$$

The fourth term is derived from the thermal noise of the resistors used in the loop filter. An example of a fourth order active loop filter is shown in Fig. 5.15.

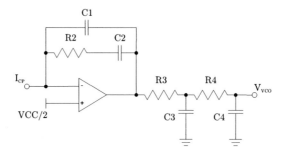

Figure 5.15: Block diagram of a 4th order active loop filter. The inverting input of the operational amplifier is fed by the charge pump output current. The non-inverting input is set to be biased with half of the charge pump supply voltage.

The corresponding noise power densities V_{Rn} of the three resistors can be calculated by

$$V_{R2} = \sqrt{4kT\ R2} \tag{5.39}$$
$$V_{R3} = \sqrt{4kT\ R3} \tag{5.40}$$
$$V_{R4} = \sqrt{4kT\ R4} \tag{5.41}$$

with k as the Boltzmann constant and T as the temperature.

The corresponding transfer functions $T(s)$ are given with:

$$T_{R2}(s) = \frac{1}{s^2 C3\, C4\, R3\, R4 + s\,(C4\, R4 + R3\, C3 + R3\, C4) + 1}$$
$$\cdot \frac{s\, C2}{s^2 C1\, C2\, R2 + s\,(C1 + C2)} \frac{1}{1 + G(s)\, H} k_{VCO} \tag{5.42}$$

$$T_{R3}(s) = \frac{1}{s^2 C3\, C4\, R3\, R4 + s\,(C4\, R4 + R3\, C3 + R3\, C4) + 1}$$
$$\cdot \frac{1}{1 + G(s)\, H} k_{VCO} \tag{5.43}$$

$$T_{R4}(s) = \frac{s\, R3\, C3 + 1}{s^2 C3\, C4\, R3\, R4 + s\,(C4\, R4 + R3\, C3 + R3\, C4) + 1}$$
$$\cdot \frac{1}{1 + G(s)\, H} k_{VCO}. \tag{5.44}$$

The contribution of each resistor to the total phase noise can be calculated by

$$L_{R2}(f) = 20 \log\left(\frac{\sqrt{2}\, V_{R2}\, |T_{R2}(j2\pi f)|}{2f} \right) \tag{5.45}$$

$$L_{R3}(f) = 20 \log\left(\frac{\sqrt{2}\, V_{R3}\, |T_{R3}(j2\pi f)|}{2f} \right) \tag{5.46}$$

$$L_{R4}(f) = 20 \log\left(\frac{\sqrt{2}\, V_{R4}\, |T_{R4}(j2\pi f)|}{2f} \right). \tag{5.47}$$

From Eq. 5.36 some important parameters can be derived. The root mean square value of the phase noise RMS_{phase} and the root mean square value of the frequency error RMS_{freq} can be calculated by

$$\text{RMS}_{phase} = \frac{180°}{\pi} \sqrt{2 \int 10^{\frac{L_{total}(f)}{10}} \mathrm{d}f} \tag{5.48}$$

$$\text{RMS}_{freq} = \sqrt{2 \int 10^{\frac{L_{total}(f)}{10}} \mathrm{d}f} \tag{5.49}$$

The error vector magnitude EVM, the jitter T_j, and the signal to noise ratio SNR are given with

$$\text{EVM} = \frac{\pi}{180°} \text{RMS}_{phase} \tag{5.50}$$

$$T_j = \frac{1}{f_{out}} \frac{\text{RMS}_{phase}}{360°} \tag{5.51}$$

$$\text{SNR} = \frac{1}{2 \int 10^{\frac{L_{total}(f)}{10}} \mathrm{d}f}, \tag{5.52}$$

where the integration has to be made over the whole calculated frequency range. The low frequency has to be as close as possible to the carrier and the upper one out of the bandwidth of the system. The simulations show that the phase noise is strongly dependent on the bandwidth and the phase margin of the closed loop. They also show that there is an optimum pair of values, where the total phase noise has a global minimum. For the presented PLL architecture this is the case for a loop bandwidth of $f_c = 110\,\text{kHz}$ and a phase margin of $\Phi = 77°$. Fig. 5.16 shows the simulation result compared to a phase noise measurement at an output frequency of $f_{out} = 7.5\,\text{GHz}$.

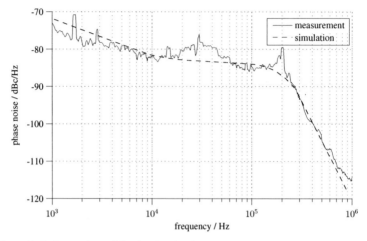

Figure 5.16: Comparison of the simulated and measured phase noise of the implemented synthesizer. The output frequency is set to be $f_{out} = 7.5\,\text{GHz}$.

The loop bandwidth for the simulation and the measurement setup was chosen to be $f_c = 110\,\text{kHz}$. It can be seen that the simulation gives only a very rough estimation of the real phase noise behavior. The reason for this is mainly the much more complex system of the demonstrator along with some non-idealities that are not respected in the phase noise model.

Another very important parameter of the synthesizer is the linearity of the generated frequency ramps. This parameter is mainly determined by the settling time of the closed loop. The modeling of the closed loop is described in appendix B in detail. Using these developed models of the closed loop, a system simulation is set up using Matlab Simulink to investigate the linearity of the frequency ramps. The corresponding measurement is done in time domain. Fig. 5.17 shows the according measurement setup. The RF signal is downconverted with a mixer and a local oscillator (LO) frequency of 6.9 GHz. Afterwards the signal is sampled using a sampling oscilloscope and a sampling frequency of 2.5 GHz. The recorded signal is then analyzed performing a short-time Fourier transform. The result is shown as a spectrogram in Fig. 5.18. In Fig. 5.19 the maxima of the spectrogram at every point in time is compared with an ideal triangular signal. It can be seen that the measured signal significantly differs from the ideal one, especially at the turning points at the end of the up or down ramps. One effect that can be observed is that the signal, after reaching the upper frequency limit of the ramp, does not immediately start the down ramp, but stays constant in frequency for a short time . These plateaus cause a slightly longer periodic time of the frequency ramps leading to a static frequency error. This effect can be eliminated by a proper programming of the DDS.

In addition to that, a non-linearity at the turning points can be seen, which can be explained by the transient effect. Fig. 5.20 shows the frequency difference of the generated signal from the ideal one at the lower turning point in measurement and simulation. A good agreement between measurement and simulation can be observed both in the duration of the distortion and in the shape of the curve. This frequency error has to be minimized since it directly transfers into the IF spectrum of the local positioning system [114]. It is strongly dependent on the design of the loop filter and the closed loop settling time. In general it can be stated that a high loop bandwidth leads to a short settling time and an enhanced linearity of the frequency ramps.

The performance of the later ranging and positioning system is mainly affected by two effects of the radar signal synthesizer: the phase noise and the linearity of the frequency ramps. In chapter 6 it will be shown that the system accuracy depends on the effective signal-to-noise ratio (SNR) of the low-pass filtered IF signal. As Roehr showed in his work the maximum achievable SNR is determined by the phase noise of the radar signal [110]. The non-linearity of the FMCW signal has a minor impact on the overall system performance as it is shown by Mosshammer [114]. However, especially when applying a very large sweep bandwidth this effect has to be accounted for and to be minimized. The investigations on the synthesizer performance show that the loop filter is the most critical part in the PLL architecture. A compromise between linearity and phase noise of the output signal has to be found to result in a good performance for the positioning system. For the implemented hardware demonstrator the loop filter is designed to result in a loop bandwidth of $f_c = 500\,\text{kHz}$ with a phase margin of $\Phi = 65°$ which leads to a fairly short settling time of the PLL and a phase noise that is always below -70 dBc.

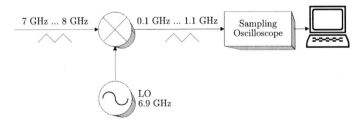

Figure 5.17: Measurement setup for evaluating the linearity of the linear frequency modulated signal. The frequency ramps with a bandwidth of $B = 1\,\text{GHz}$ and a center frequency of $f_c = 7.5\,\text{GHz}$ are mixed down with a local oscillator with a frequency of $f_{lo} = 6.9\,\text{GHz}$. The downconverted signal is stored with a digital sampling oscilloscope and post-processed using Matlab.

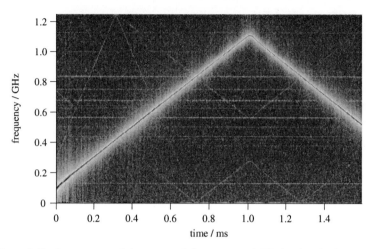

Figure 5.18: Spectrogram of the measured downconverted RF signal. The frequency ramp starts at 7 GHz and the sweep bandwidth is programmed to be 1 GHz. After the 8 GHz are reached, the frequency of the output signals stays a short time constant before it starts to decrease, causing a small plateau and slightly increasing the periodic time of the frequency ramps.

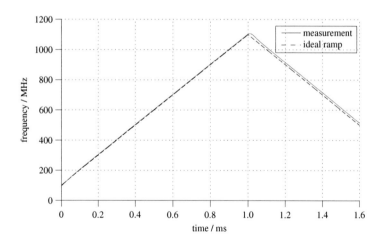

Figure 5.19: Comparison of the measured frequency ramp and an ideal ramp. A significant deviation at the turning point can be observed.

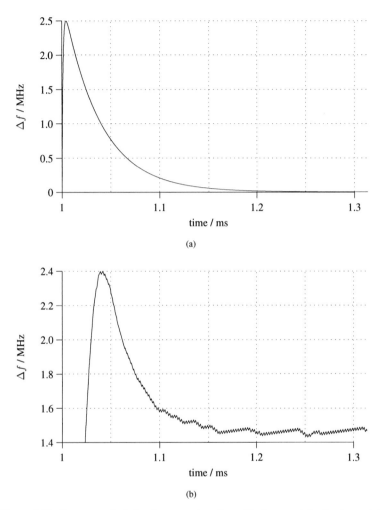

(a)

(b)

Figure 5.20: Frequency deviation in comparison to an ideal ramp at the turning point of the frequency ramp. Simulation (a), measurement (b). A good matching between simulation and measurement is obtained. The reason for the measured deviation not to drop back to zero is that the implemented synthesizer stays a short time at the upper frequency before the down ramp begins, causing a constant frequency error compared to the ideal ramp.

The Chopper

As described in section 5.3, the generated frequency modulated RF signals have to be chopped to get in compliance with the international regulation standards. To solve this issue, a dedicated fast RF switch, controlled by a pulse generator, is inserted in the transmission path as depicted in Fig. 5.14. For the PFM-UWB system, the generation of very narrow pulses is critical for the overall system performance. A comparison between different pulse generator techniques is given in appendix C. To demonstrate the functionality of the PFM technique, a signal synthesizer is implemented with full compliance to the European UWB regulations. For this reason the widening of the transmission spectrum is not a critical issue compared to a FCC compliant system. However the shown principle can easily be adapted according to the U.S. regulations by using shorter pulses for the chopping of the FMCW signals. For the described system, the chopper is designed to realize a pulse width to pulse period ratio of approximately $1/12$ with a pulse width of $t_{pw} = 3.5\,\text{ns}$. For this purpose a standard commercial available single-pole double-throw (SPDT) RF switch is used. The most important switch parameters as provided by the manufacturer datasheet are summarized in table 5.2 [115]. The pulse is generated by a self-resetting D flip-flop which is triggered by a 40 MHz sinusoidal signal. The system is in full compliance with the European regulations given by the European Telecommunications Standards Institute (ETSI). Fig. 5.21 shows the result of the measurement of the average power spectral densities without and with the chopping. The measurement is performed according to the measurement technique described in appendix A. The signal synthesizer is programmed to generate frequency ramps between the frequency $f_l = 7\,\text{GHz}$ and $f_u = 8\,\text{GHz}$ with a CW power of 0 dBm. The spectrum without the chopping shows an average power spectral density (PSD) of approximately -30 dBm/MHz as it is expected from the theory presented in section 5.3. The decrease of the average signal power around the frequency of 7.6 GHz is caused by the output characteristics of the used VCO. By inserting the RF switch to chop the signal with an on-off ratio of approximately $1/12$, the average PSD is decreased by approximately 12 dB and the -41.3 dBm/MHz UWB limit is met.

System Parameter	Value
Insertion Loss	1.6 dB
Frequency Range	DC - 50 GHz
Input Return Loss	15 dB
Output Return Loss	15 dB
Switching Speed:	
10%-90& rise time	100 ps
90%-10& fall time	90 ps

Table 5.2: Parameters of the used RF switch for chopping the FMCW signal.

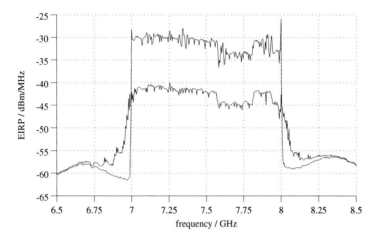

Figure 5.21: Measured EIRP of a FMCW signal without and with chopper. Without the chopping of the signal, the average PSD is given with -30 dBm/MHz. By inserting a fast RF switch which is triggered by 3.5 ns pulses and a pulse repetition frequency of 40 MHz the average PSD is decreased and drops below the -41.3 dBm/MHz limit.

5.5.2 The Receiver Chain

In Fig. 5.22 a block diagram of the receiver chain of the implemented demonstrator system is depicted. The UWB signal, received by the antenna, is fed to a first amplifier stage by a duplexer, consisting of a simple single-pole, double-throw switch. The signal is then filtered by a band-pass filter structure, which is composed of a microstrip low-pass filter, followed by a surface mounted high-pass filter. The resulting filter characteristic is depicted in Fig. 5.23. After limiting the bandwidth with the filter, the received signal is again amplified by a second low-noise amplifier (LNA) and fed to a mixer. After the downconversion, the signal is fed to an active low-pass filter with a cut-off frequency of $f_c = 2.0\,\text{MHz}$. The filter also transfers the single-ended signal to a differential signal. After the filtering, the IF-signal is sampled by an analog-to-digital converter (ADC) and the digital data is processed by a DSP. The parameters of the used RF components are given below the corresponding symbols in Fig. 5.22. Using the well known Friis formula, it is possible to calculate the total noise figure F_{total} of the complete receiver chain. It is defined by

$$F_{total} = F_1 + \frac{F_2 - 1}{G_1} + \frac{F_3 - 1}{G_1 G_2} + \frac{F_4 - 1}{G_1 G_2 G_3} + \cdots. \tag{5.53}$$

Using the figures depicted in Fig. 5.22 with equation 5.53 a theoretical total noise figure of 3.1 dB is achieved for the total receiver chain. Hence, for a wanted SNR of 20 dB at the ADC input, a minimum SNR at the duplexer input of 23.1 dB is needed. Assuming a temperature of 290 K, the theoretical input noise density can be calculated using the Boltzmann constant $k = 1.3810^{-23}$ J/K resulting in $P_{noise} = -174.0\,\text{dBm/Hz}$.

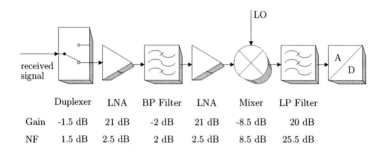

	Duplexer	LNA	BP Filter	LNA	Mixer	LP Filter
Gain	-1.5 dB	21 dB	-2 dB	21 dB	-8.5 dB	20 dB
NF	1.5 dB	2.5 dB	2 dB	2.5 dB	8.5 dB	25.5 dB

Figure 5.22: Block diagram of the receiver chain of the implemented hardware demonstrator and the according component parameters. The received signal is fed to a first low-noise amplifier. Afterwards it is filtered by a band-pass filter structure and again amplified by a second low-noise amplifier. It is then downconverted, filtered by an active low-pass filter and converted to a digital signal by an analog-to-digital converter.

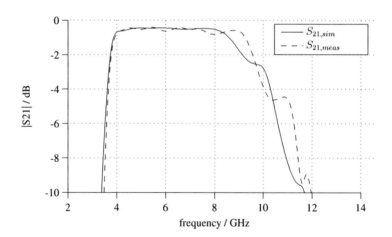

Figure 5.23: Simulated and measured S_{21} of the implemented band-pass structure. The filter is realized by microstripline low-pass filter followed by a surface mounted high-pass filter. A good agreement between measurement and simulation can be observed.

5.5.3   UWB Antennas

In conventional narrowband systems, antennas are not such a critical element in the signal flow as in UWB systems, where the antenna acts as a filter for the transmitted and received UWB signal.

In general an antenna is a device that transmits and / or receives electromagnetic waves. The transmission and reception of the electromagnetic waves should be nearly reflection free. Thus, the antenna has to transform the characteristic impedance of the frontend circuit Z_0 to the free space wave impedance Z_{F0} with

$$Z_{F0} = \frac{\vec{E}}{\vec{H}} = \sqrt{\frac{\mu_0}{\epsilon_0}} = 120\pi \ \Omega. \tag{5.54}$$

This transformation can be done by shape and size of the antenna. Traditional UWB antennas are typically multi-narrowband antennas instead of antennas optimized to receive a single coherent signal across their bandwidth. An UWB system requires an antenna capable of working on all frequencies at the same time [116]. Thus, antenna behavior must be consistent and predictable across the entire band. Ideally, pattern and matching should be stable across the entire band. For conventional antennas, the reciprocity theorem can be applied, meaning that the antenna has the same receiving and transmitting characteristics. For the PFM-UWB system demonstrator, four different antennas are investigated. The first one is an off-the-shelf UWB antenna *SMT-3TO10M-A*, provided by the U.S. company *Skycross*. The second antenna under investigation is a coplanar monopole structure designed at the University of Karlsruhe, Germany. The presented third antenna is a monopole ground planar structure, which shows good performance in regard of omnidirectivity. The last antenna is an ultra low-cost design using two two-cent coins to form an elliptical dipole antenna. In practice the monopole over ground planar antenna shows the best overall performance for the mobile stations, while the highly directive coplanar monopole antenna were ideally suited for the base station units.

Internal 3-10 GHz stamped metal antenna by Skycross

In Fig. 5.24a the *SMT-3TO10M-A* UWB antenna by *Skycross* is shown [100]. The antenna consists of a bended sheet as the antenna element, which is placed on a PCB with a female SMA connector. Fig. 5.24b depicts the measured return loss of the antenna. It can be seen that the *SMT-3TO10M-A* performs well in the desired frequency region. The radiation pattern for the azimuth is almost omni-directional and the overall size of the antenna is 2.62 cm x 1.85 cm.

UWB Coplanar Monopole Antenna

In Fig. 5.25a a very compact implementation of a coplanar monopole antenna is shown, designed by the University of Karlsruhe, Germany [117]. The radiating element is directly etched on the ground plane and is fed by a coupling feeding network placed on the backside of the dielectric substrate consisting of FR4 with $\epsilon_r = 4.5$ and a thickness of 1.575 mm. The antenna is very compact with an overall size of 2.3 cm x 3 cm.

In Fig. 5.25b the measured return loss ($|S_{11}|$) is illustrated. It can be seen that the antenna structure has a good matching over the whole desired frequency range. Fig. 5.25c shows the measured gain in the H-plane. The antenna presents a main lobe at 0 degrees, extended in the angular region -50 / +50 degrees. The gain in the main lobe is rather high, forming a quite directive antenna. Pancera et al. show that the structure provides a smooth transition from the transmission line to the free space. Hence, the proposed antenna has a good time domain behavior. In fact, since the impulse response is very short, the distortion applied by the antenna to the transmitted signal is low.

UWB Monopole Over Ground Planar Antenna

Since an omnidirectional antenna characteristic is very favorable for the mobile units, an easy and cheap antenna solution had to be found. Fig. 5.26a shows an implemented low-cost monopole over ground planar antenna. The antenna offers a compact form factor while still supporting a wide range of frequency bands, continuously. The planar structure is implemented on a Rogers 4003 substrate with $\epsilon_r = 3.38$ and a thickness of 0.51 mm. The particular shape of this antenna is chosen to be able to parameterize the fundamental radiating structure in order to determine the critical variables that influence the behavior of the antenna. A spline-based shape is chosen and built. Starting from the feed-point, a progressive tapering of the transmission line takes place, leading smoothly to a corner-less radiating element that allows the wave to be radiated. The spline shape is chosen to build the antenna since it allows to easily parameterize such a smooth element with well defined dimensions. A monopole-over ground topology is chosen, to ease the feeding of the antenna, allowing low cost implementation and hassle free integration onto RF circuit boards. Fig. 5.26b shows the measured return loss ($|S_{11}|$) of the planar antenna structure. The covered frequency range is from 5 GHz to 9 GHz, the radiation pattern is azimuth omnidirectional with linear polarization and the feed impedance is matched to 50 Ω. The peak gain is at 7.5 GHz with 4.2 dBi. It can therefore be inferred that with this antenna, it is possible to work in the desired UWB frequency band, as well as in the 5.8 GHz ISM band. An illustration of the directivity of the antenna is given in Fig. 5.26c.

Four-Cent Antenna

UWB antennas need not to be complex or expensive as shown in the following imple-mented antenna design. Fig. 5.27a shows an implemented elliptical element dipole using two Euro two-cent pieces. The antenna performance is mainly specified by two charac-teristics of the structure. The diameter of the copper coins affects the lower bandwidth frequency, while the thickness and the gap between the two elements affects the match. In Fig. 5.27b the return loss ($|S_{11}|$)of the Four-Cent antenna, measured with a VNA, is depicted. The gain of these antennas is between 0 dBi and 3 dBi in the frequency range from 5 GHz to 9 GHz. A more complete analysis of this kind of UWB antennas can be found in [118].

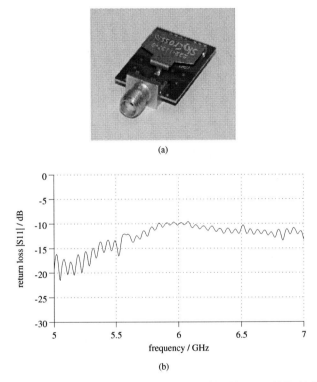

Figure 5.24: UWB antenna *SMT-3TO10M-A* fabricated by *Skycross* [100]. (a) Picture of the antenna, (b) measured return loss.

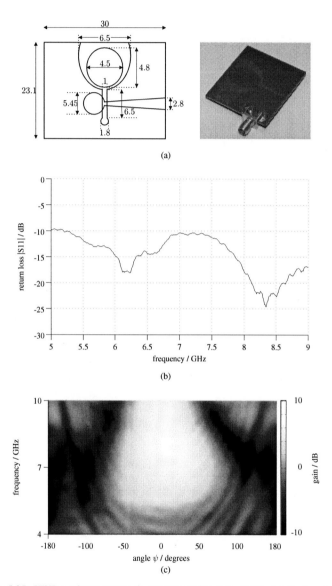

Figure 5.25: UWB coplanar monopole antenna designed by [117]. (a) Design drawing and picture of the antenna (dimensions in millimeters), (b) measured return loss, (c) measured antenna gain in the H-plane for co-polarization as a function of the angle ψ.

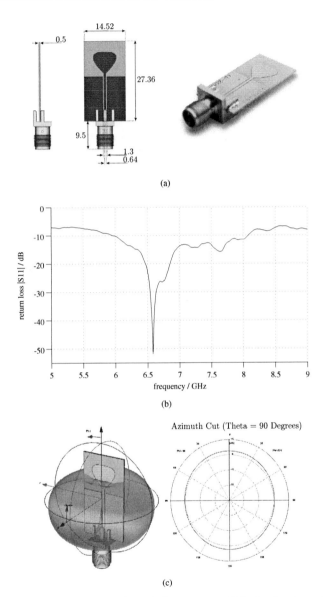

Figure 5.26: UWB monopole over ground planar antenna. (a) Design drawing and picture of the antenna (dimensions in millimeters), (b) measured return loss, (c) farfield plot and azimuth cut of the antenna at 7.5 GHz.

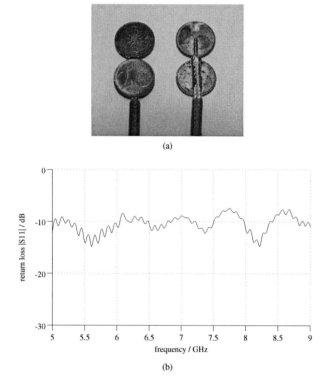

(a)

(b)

Figure 5.27: Implemented very low-cost four-cent antenna. (a) Picture of two implemented elliptical element dipoles *Austria* and *Netherlands*, (b) measured return loss.

5.5.4   The Prototype System

After successful system simulations and evaluation of all single system modules a prototype measurement system was implemented. The complete system can be subdivided into four major parts:

- Power supply

- Digital module

- IF module

- RF module

where the first three parts were developed by the *Symeo GmbH*. As mentioned before, the complete system control is done by a dedicated DSP. The DSP is the central processing unit and performs the calculations for synchronization, distance measurement and the overall system control. It also programs the DDS which generates the reference signal for the PLL. The RF module contains the radar signal synthesizer, the chopper, and the receiver chain from the duplexer to the output of the active low-pass filter. Fig. 5.28 depicts a close up of the RF module. The key components that have been discussed in the previous sections in detail are labeled in the figure. The downconverted and filtered signal is then fed to the IF board, where the signal is digitized by an ADC. The IF board also contains the DDS providing the reference signal for the PLL on the RF board. The DSP and all other system parts are clocked by a single crystal oscillator. The hardware for base stations and mobile units is completely identical. Depending on the software settings, the modules either work as base stations or mobile units. The prototype can be supplied by a optional battery pack, which allows the complete mobility of the mobile units. Fig. 5.29 shows the printed circuit board (PCB) stack of the prototype system and the respective dimensions.

For the prototype system it is also possible to connect an inertial measurement unit (IMU). This additional hardware module, developed by the University of Clausthal, Germany, contains several gyroscopes and 3D acceleration sensors. The data obtained from these sensors can be used to further increase the accuracy of the positioning system by performing a sensor fusion and in addition to that it can be used to maintain the localization ability of the unit even in absence of a LOS for the radar signal. In Fig. 5.30 a set of several measurement units as they were used for the performance measurements are shown.

For the last distance measurement campaign the RF board was slightly modified. Instead of the usage of the chopping principle to decrease the average transmission, power a simple programmable signal attenuator was used to simplify matters. The modification does not affect the principle function and takes the last developments of the European regulation authorities into account which don't exclude pure FMCW-UWB systems anymore.

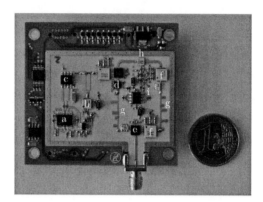

Figure 5.28: Picture of the implemented RF module. The main parts of the module: (a) PLL IC, (b) VCO, (c) PLL prescaler, (d) RF chopper, (e) duplexer, (f) LNA, (g) band-pass structure, (h) mixer, (i) active low-pass filter.

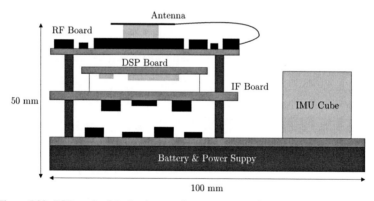

Figure 5.29: PCB stack of the implemented measurement unit prototypes. The complete measurement modules consist of an optional battery and a power supply and distribution PCB, the IF Board, the DSP circuit board, and the RF board with the antenna. In addition to that an IMU can be used together with the radar positioning system to further enhance the accuracy of the positioning system and to maintain the localization ability even in absence of a LOS for the RF signals.

(a)

(b)

Figure 5.30: Pictures of prototype modules as they were used for the performance measurements.

CHAPTER 6

System Evaluation and Measurement Results

To characterize the system performance of the previously described local positioning system, various measurement campaigns have been performed. Four different setups are chosen to evaluate the performance of the developed PFM UWB positioning system:

- Measurement of the length of a delay line

- Distance measurement in outdoor environment

- Distance measurement in indoor environment

- Distance measurement with a linear unit in indoor environment

The first setup is used to investigate the best case performance of the measurement system without the influence of multipath effects. Since in this setup, the RF signals can be attenuated very easily, the influence of the SNR on the system performance is evaluated by inserting a step attenuator in the delay line. With these results, a rough estimation on the maximum operating range is obtained.

Subsequently the measurement system is tested in an outdoor environment, where the influence of the environment is at a minimum. Furthermore in outdoor environment the maximum operation range is tested. The measurement results of the PFM-UWB system are compared to the results of an industrial laser ranging system.

Finally the system is evaluated in an indoor environment. In the initial indoor scenario the system is tested in a hallway of an office building and the laser measurement system is used to verify the measurement results. The second indoor measurement setup uses a linear unit as reference system. While the base station is located at a fixed position, the mobile unit is mounted on an automatic sledge with high positioning accuracy, allowing an even more accurate reference measurement without the uncertainties introduced by the laser reference system. For all measurements a sweep bandwidth of $B = 1\,\text{GHz}$ and a sweep duration of $T = 1\,\text{ms}$ is used.

6.1   Delay Line Measurements

To evaluate the best case performance of the measurement system, a base station and a mobile unit of the PFM-UWB system are connected by a delay line. For this purpose a coax cable of type *LL2773-AF* [119] is chosen. The physical length of the coax cable is approximately 100 m. Since the length of the delay line is measured by a microwave signal, propagating through the coax cable, the obtained measurement result gives the electrical length of the delay line. This length depends on the mechanical length and the relative permittivity of the used coax cable and can be assumed as 120 m for the used *LL2773-AF*. Fig. 6.1 shows the measurement setup for the delay line measurement. In addition to the length of the delay line, the measured length is increased by approximately 20 m because of additional coax cables, used to connect the units with the delay line. Therefore a mean value between 140 m and 150 m is expected. As already mentioned, this measurement setup represents the best case scenario for the measurement system, since no distortions caused by multipath signal propagation are introduced. Furthermore the attenuation of the RF signals is known exactly.

The distribution of the measured length of the delay line is shown in Fig. 6.2. The probability density function (PDF) of the distance measurement results is obtained from 2500 measurement samples of the electrical length of the delay line. The measured distances (black line) are compared to the probability function of a Gaussian random variable with the same mean value and standard deviation (grey line). A good agreement between both distributions is observed which implies that the distance measurements are random samples from a Gaussian distribution. The standard deviation for the measured length is 6.57 mm. Given the system bandwidth of $B = 1$ GHz and the sweep duration of $T = 1$ ms, the standard deviation of 6.57 mm implies a synchronization of both stations to an offset in time and frequency of less than 44 ps and 44 Hz respectively.

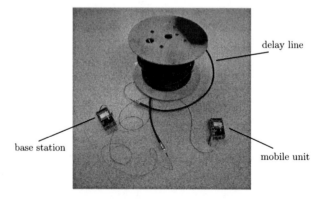

Figure 6.1: Setup of the delay line measurement. A base station is connected to a mobile unit via a coax cable. By using the PFM UWB measurement system, the electrical length of the delay line is measured.

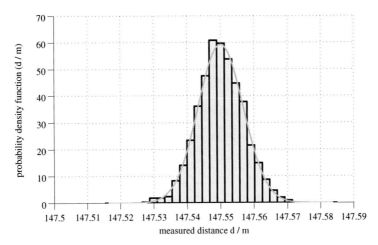

Figure 6.2: Distribution of distance measurements of a delay lines. 2500 measurement samples are used to calculate the probability density function, which approximates a Gaussian distribution. The according standard deviation is below 7 mm.

To observe the performance of the system in dependence on the level of the received signal, the two units are again connected via a short delay line and a step attenuator is inserted in between the transmission path. The attenuation can be set to values from 0 dB to 57 dB. The attenuation of the coax cable is approximately 31 dB for the used frequency range. In total a range from 31 dB to 85 dB of attenuation is covered were a successful synchronization and distance measurement could be performed. The step attenuator is initially set to 0 dB of attenuation and increased by steps of 3 dB. For each attenuation a set of 650 samples of the measured delay line length is acquired and evaluated. Fig. 6.3 shows the variation of the measured mean value over the applied attenuation. The mean values are within ±4 mm from their average and therefore it can be stated that the mean value is independent of the signal attenuation, as it does not change significantly over all applied attenuator steps.

Subsequently Fig. 6.4 shows the calculated standard deviation of the distance measurement results over the respective signal attenuation. It can be seen that, as long as the total attenuation does not exceed 73 dB, the standard deviation is below 7 mm. Beyond an attenuation of 73 dB the standard deviation increases up to a maximum of 12.6 mm at a total signal attenuation of 85 dB. If the attenuation exceeds 85 dB the level of peaks in the power spectrum densities of the low-pass filtered IF signal is too low to be detected and therefore a synchronization and distance measurement is not possible anymore. With the obtained maximum allowable signal attenuation, a rough estimation of the maximum operating range of the signal can be calculated using the free-space path loss formula solved for the distance d and using the signal attenuation a_{path}:

$$d = \frac{\lambda}{4\pi} 10^{\frac{a_{path}}{20}}, \tag{6.1}$$

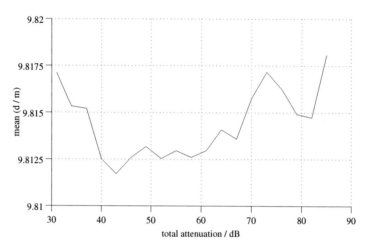

Figure 6.3: Mean value of distance measurements over applied signal attenuation. The total attenuation is given as the sum of the attenuation of the coax cable (31 dB) and the attenuation of the step attenuator from 0 dB to 57 dB. The total covered attenuation range is therefore from 31 dB to 88 dB. The mean value does not change significantly.

assuming the use of omni-directional antennas with no directional gain. Using the center frequency $f_c = 7.5\,\text{GHz}$ a signal attenuation of 73 dB corresponds to a transmission range of 15 m where a low standard deviation is achieved. The maximum attenuation of 85 dB corresponds to a maximum operating range of 57 m. However, even at this possible maximum distance a remarkable precision of the measurement system is maintained.

As Roehr already derived in his work [110], up to a certain attenuation (here 73 dB), the effective SNR near the peak in the power spectral density of the IF signal of the PFM-UWB system is, as well as in the *Symeo* LPR, only determined by the phase noise of the implemented PLL. In Fig. 6.5 two exemplary power spectral densities are shown, obtained during the delay line measurements without any attenuation in addition to the loss in the coax cable and with a supplemental attenuation of 40 dB. For the first measurement setup (grey line) the noise plateau caused by the phase noise of the PLL is clearly above the noise floor and defines the effective SNR. By adding additional attenuation in the signal path, the peak power corresponding to the LOS signal drops, as well as the phase noise plateau, so that the effective SNR stays approximately constant allowing a constant and low standard deviation. If the level of the plateau drops down to the thermal noise floor of the system (black line), this noise floor becomes the limiting factor for the effective SNR. By further increasing the signal attenuation, the noise floor remains constant and is determined by the thermal noise of the receiver. Only the level of the peak decreases causing a reduction of the effective SNR, leading to an increased standard deviation in the distance measurement. If a total attenuation of 85 dB is reached, the SNR becomes too small to allow a reliable detection of the peak.

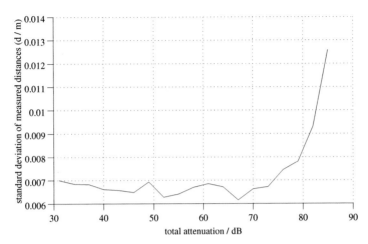

Figure 6.4: Standard deviation of acquired distance measurements over applied total signal attenuation. The standard deviation is below 7 mm as long as the total attenuation is below 73 dB. Afterwards the standard deviation increases as the attenuation increases as well. The maximum attenuation where a successful synchronization and distance measurement can be performed is 85 dB with an according standard deviation of 12.6 mm.

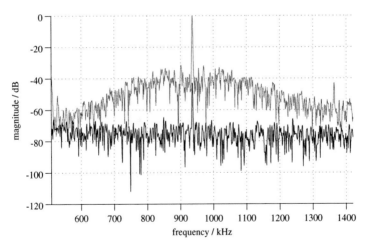

Figure 6.5: Two exemplary power spectra obtained during the delay line measurements. For low attenuation (grey line) the noise plateau caused by the phase noise of the PLL defines the effective SNR. As the attenuation increases, both, the LOS peak power as well as the phase noise plateau, decrease (black line).

6.2   Outdoor Distance Measurements

After the evaluation of the best-case performance of the PFM-UWB system in the previous section, the system is setup in an outdoor location next to an office building. Fig. 6.6 shows the measurement setup. One unit, acting as the base station is placed on a fixed position, while the second unit, configured as a mobile unit is mounted on a trolley. In this scenario it is expected that only reflections from the ground and from the office building will degrade the performance of the system. An industrial laser ranging system serves as reference distance measurement system to validate the measurement results obtained by the PFM-UWB system. Since there is an offset between the laser ranging system and the distances measured by the PFM-UWB radar caused by the module arrangement the system has to be calibrated before the measurements are done. To calibrate the system the offset between the laser ranging system and the microwave ranging system is set to zero at a certain distance. Both radar measurement units are each connected to a coplanar monopole antenna presented in section 5.5.3. The directional antennas are chosen to further reduce the influence of the signal reflections caused by the office building next to the measurement setup. To stay in compliance with the UWB legal requirements, the used maximum transmit signal power is adjusted in a way that the directional gain of approximately 14 dBi of the antennas is accounted for, which means the peak power has to be reduced by 14 dB. However, the maximum operating distance of the system is increased because of the antenna gain for the receiver, causing an increased received signal level. The maximum distance, the system is tested is 72 m.

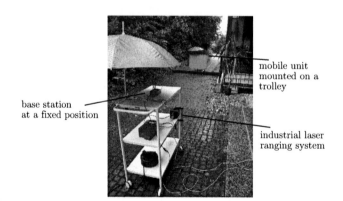

Figure 6.6: Setup of the distance measurement at an outdoor environment. A base station is placed at a fixed position while the mobile unit is mounted on a trolley. An industrial laser ranging system is used as reference measurement system to compare the acquired radar distance values. For the measurement campaign, the trolley is set up at the far end of the measurement field at a distance of 72 m and moved towards the base station.

An analysis of the received signal power of the measurement receiver over the distance shows that at a distance of 20 m the level of the received signal drops significantly and increases again afterwards. Obviously at this certain position the signals of the direct LOS signal and the various reflections interfere with each other in a destructive way. While the trolley is moved from the maximum distance towards the base station the measurement values of the radar system, as well as the distance values from the laser ranging system are monitored continuously. In Fig. 6.7 the acquired distance measurement samples are presented, where the trolley is continuously moved towards the base station. The results of this first scenario show a good performance of the system with no principal errors and a good real time ability.

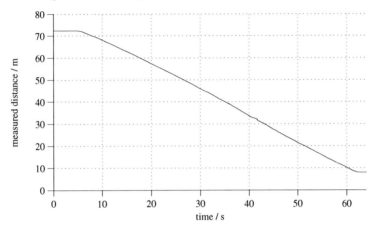

Figure 6.7: Outdoor distance measurement. The mobile unit is mounted on a trolley moving steadily from a maximum distance of 72 m towards the base station.

To observe the system more accurately the trolley is again placed at the maximum distance of 72 m and moved towards the base station in steps of 4 m. At every position 1500 measurement samples are recorded before the trolley is moved to the next position. In Fig. 6.8 the difference between the mean values of the measurement samples of the radar system and the measured distances of the laser ranging system is depicted. The mean difference is in the centimeter range with a maximum of 10.5 cm at a distance of 20 m. As showed before at this certain distance the level of the received signal drops significantly and therefore causing the large error in the mean deviation. Fig. 6.9 depicts the standard deviation of the distance measurement samples over the measured distances of the laser ranging system. The standard deviation at all distances is below 8 mm, except for the distance around 20 m, where the signal peak power level drops below -73 dB causing an increased standard deviation because of the decreased effective SNR as shown in section 6.1. The measurement results presented in the previous section also imply that the low standard deviation is expected for distances up to 71 m if antennas with a directional gain of 14 dBi are used. This calculations are in accordance to the measurement result shown in Fig. 6.9.

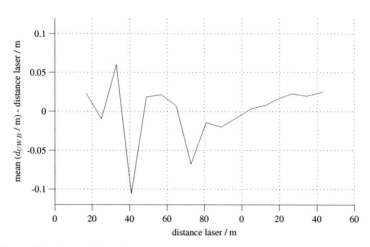

Figure 6.8: Mean deviation of the distance measurements of the radar and the laser ranging system over the distance. The systems performs well over the entire operating range but shows a significant increase in the absolute error at a distance of approximately 20 m.

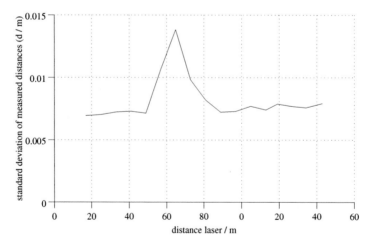

Figure 6.9: Standard deviation of the distance measurement values over the measured distance of the laser ranging system. A standard deviation of less than 10 mm over the entire measured range is obtained except around the distance of 20 m, where the received signal level drops, causing an increased standard deviation.

6.3   Indoor Distance Measurements

In the previous section the system has been evaluated in an outdoor environment with only moderate interference caused by multipath signal propagation. Next the measurement system is setup in an office building to evaluate the performance in indoor environment. In Fig. 6.10 the measurement setup located in a hallway of an office building is shown. The base station is set up at a fixed position mounted on a table along with the industrial laser ranging system already used in the outdoor measurements. Both radar stations are connected to an omni-directional antenna presented in section 5.5.3. Because of the narrow corridor and the usage of omni-directional antennas, this setup presents a challenging environment for the UWB radar system since many multipath signals are expected to be received by the antenna. The trolley is placed at the opposite end of the corridor at a maximum distance of 33 m and continuously moved towards the table with the base station. Fig. 6.11 depicts the acquired distance samples of the measurement system during this initial indoor scenario. It can be seen that the system still performs very well in in this challenging environment with no major failures. For a deeper analysis, the trolley with the mobile unit is again placed at the end of the corridor and moved towards the base station in steps of approximately 2 m. At every position 1500 measurement samples are recorded before the trolley is moved 2 m to the next position. In Fig. 6.12 the cumulative distribution function of the total error is displayed. Here the probability of a certain accuracy is shown over the total error, defined as the difference between the laser ranging system and the radar measurement system. From this plot it can be extracted, that in 70% of all measurements the accuracy is below 10 cm and in 91% of all cases, the total error is below 20 cm.

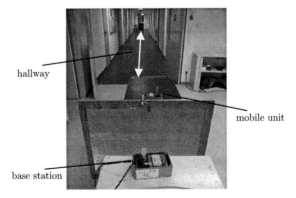

Figure 6.10: Setup of the distance measurement at an indoor environment. The measurement system is setup in hallway of an office building. The base station is placed at a fixed position, while the mobile unit is mounted on a trolley, which is moved along the hallway. An industrial laser ranging system serves as reference distance measurement system.

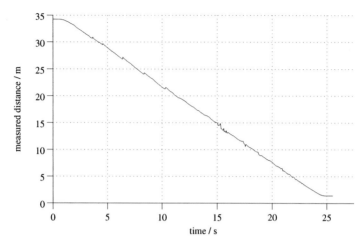

Figure 6.11: Distance measurement in an indoor environment along an office hallway. The mobile unit is mounted on a trolley moving continuously from a maximum distance of 33 m towards the base station.

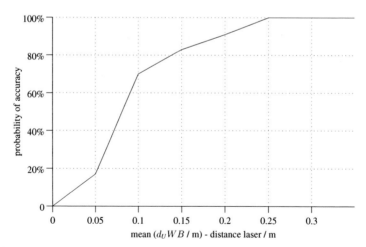

Figure 6.12: Cumulative distribution function of the total error of the radar system compared to the laser ranging system. Despite the distinctive multipath characteristic of this measurement scenario an accuracy of better than 10 cm in 70% of all cases is obtained.

The standard deviation of the distance measurements is depicted in Fig. 6.13. At all distances it is below 11 mm with a maximum of 10.7 mm at a distance of 32.1 m. The presence of multipath interferer is observed by analyzing the spectra of the low-pass filtered IF signal. An example of such a spectrum is shown in Fig. 6.14. The spectrum is recorded during the distance measurements at a distance of 20 m between the base station and the mobile unit. Besides the LOS peak located at approximately 384 kHz there are several peaks close to the LOS peak caused by the multiple signal propagation paths. The multipath characteristic follows the principles derived in chapter 4 and is the main cause for the system performance degradation. However, the experimental setup shows that the system is able to maintain its high precision and accuracy even in distinctive multipath indoor channels.

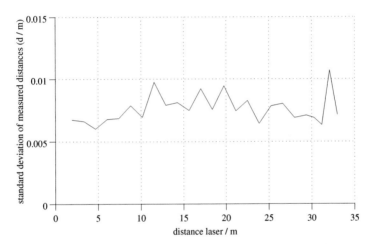

Figure 6.13: Standard deviation of the distance measurement values over the measured distance of the laser ranging system. A standard deviation of less than 11 mm over the entire measured range is obtained.

To observe the ability of the measurement system to resolve signals which follow the LOS signal after a very short time interval in the range of a few ns, an additional experimental setup is used. In this setup the two radar units are placed at fixed positions with a distance of approximately 2 m between both. Behind the mobile unit a corner reflector is set up at a distance of approximately 35 cm as shown in Fig. 6.15. With this system arrangement a defined multipath signal component is generated with a time delay of approximately 2.3 ns with respect to the RF signal propagating directly from the transmitter to the receiver. Fig. 6.16 depicts the spectrum of the IF signal acquired during a distance measurement. Besides the peak corresponding to the LOS signal at a frequency of 265.5 kHz a second peak is present at a frequency of 270.6 kHz. The difference between those frequencies $\Delta f = 5.1$ kHz correlates to a distance of 76 cm which the signal that is reflected at the corner reflector has to travel more. The results match the assumptions made for the measurement setup and demonstrate the very good ability of the system to resolve multipath signals from the LOS signal and therefore the robustness towards multipath distortions.

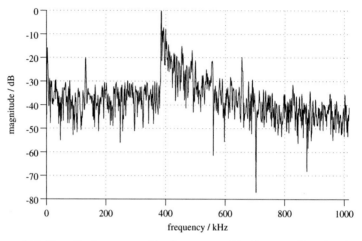

Figure 6.14: Normalized spectrum of the downconverted and low-pass filtered IF signal. The peak corresponding to the LOS signal is followed by several peaks caused by echoes.

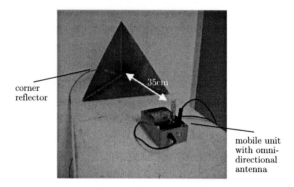

Figure 6.15: Setup for testing the robustness towards distortions caused by multipath signal propagation. Behind the unit a corner reflector is placed with a distance of approximately 35 cm between the radar antenna and the reflector. Thus a second well defined signal arrives at the receiver antenna with a time delay of approximately 2.3 ns compared to the direct signal path.

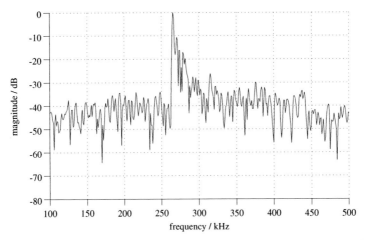

Figure 6.16: Normalized spectrum of the IF signal. Besides the first peak corresponding to the LOS signal, a strong second peak can be observed at 270.6 kHz. This peak is caused by the signal reflected at the corner reflector. The two peaks can clearly be distinguished and the short multipath component does not affect the LOS peak.

6.4   Measurements on a Linear Stage

To overcome the uncertainties which may be introduced by inaccuracies of the trolley movement or the laser ranging system as reference system, the UWB radar system is evaluated using a linear stage and an automated sledge in indoor environment. The measurement setup is shown in Fig. 6.17. Here the base station is placed on a stand at a fixed position. The mobile unit is mounted on an automatic sledge that can be moved along the linear stage. The true position of the sledge is known with an inaccuracy of less than 1 mm so that the obtained errors in the radar distance measurement can be assumed to be caused by the UWB LPR solely. The sledge is moved to one end of the linear stage at a distance of 4 m from the base station. It is the moved towards the base station in steps of 2 mm. At each position 650 samples of the distance measurement are acquired before the sledge is moved to the next position. In contrast to the measurement setup presented in the previous section the exact knowledge of the true distance between the two radar units allows for a detailed analysis of the system performance in indoor environment with an exact reference system.

The mean error in the distance measurement of the radar system over the exact position obtained from the linear stage is depicted in Fig. 6.18. The total accuracy is within ±5.2 cm over the entire distance from 0 m to 4 m. Fig. 6.19 depicts the cumulative distribution function of the error in the distance measurement of the UWB-LPR. From this plot it can bee seen that in 84% of all cases the error is below 3 cm and in 60% of all measurements the accuracy is better than 2 cm.

The standard deviation for this measurement setup is shown in Fig. 6.20. It is always below 1 cm with a mean value of 7 mm and matches the standard deviation obtained in the delay line measurement setup of section 6.1.

Figure 6.17: Measurement setup using a linear stage. The base station is placed at one end of the linear stage while the mobile unit is mounted on an automatic sledge moving from one end towards the base station in steps of 2 mm over a total distance of 4 m.

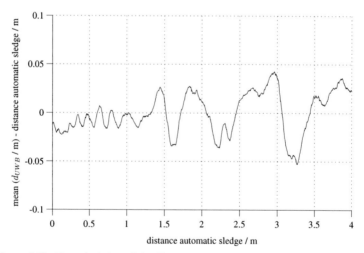

Figure 6.18: Mean deviation of the distance measurements of the UWB radar and the obtained values of the linear stage over the distance. The maximum deviation is 5.2 cm emphasizing the high accuracy the system can achieve even in indoor environments.

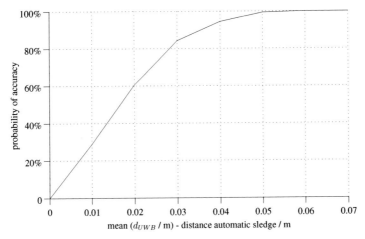

Figure 6.19: Cumulative distribution function of the total error of the radar system compared to the absolute position obtained by the linear stage. In this indoor environment with distinctive multipath characteristic an accuracy of better that 3 cm in 85% of all cases is obtained and always better than 5.2 cm.

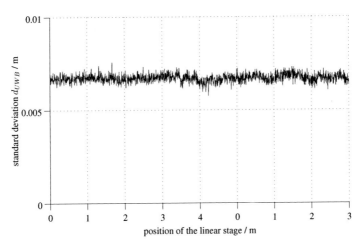

Figure 6.20: Standard deviation of the distance measurement values over the distance values obtained by the linear stage. The system shows a standard deviation of approximately 7 mm over the entire measured range.

Ceterum censeo Carthaginem
esse delendam.

Marcus Porcius Cato, Senior

CHAPTER 7

Conclusions and Outlook

This work presents the investigation and analysis of a novel pulsed frequency modulated ultra-wideband ranging and positioning system. The system is developed for the use in very harsh industrial environments.

A profound knowledge of the radio channel is crucial for the development of a wireless ranging and positioning system using RF signals. Thus, several channel aspects are discussed and measurement results are given to estimate the feature channel characteristics. A detailed analysis of indoor radio channel characteristics shows that especially in typical production environments, many metal obstacles and walls lead to a sever performance degradation of narrowband positioning systems. This is mainly induced by multipath signal propagation caused by signal reflections. These reflections superimpose the line-of-sight signal and hinder an accurate estimation of the distance between two nodes. One possibility to obtain good resolution of multipath components is the usage of a very large signal bandwidth such as with ultra-wideband signals.

The thesis at hand introduces a novel UWB technique, using a combination of the common FMCW radar approach and a pulsed ultra-wideband technique. With this method the advantages of both techniques can be utilized. The measurement system is comprised of two equal units, the base station and the mobile unit. After a high precision synchronization of the two independent nodes, the mobile sends a synchronized reply back to the base station and the distance between these two units is measured via the round-trip time-of-flight of a FMCW signal. To comply with international UWB regulations a fast RF switch is used to chop the transmit signal. A prototype has been developed using a center frequency of 7.5 GHz and a bandwidth of 1 GHz. The demonstrator system shows for the first time the feasibility of the pulsed frequency modulated UWB technique for ranging and positioning in full compliance with international UWB regulations.

A major focus of this thesis is on the evaluation of the performance of the measurement system. Therefore extensive measurements have been performed to obtain the system characteristics. The system was tested in various application scenarios in outdoor, as well as in indoor environments. The measurement results indicate a promising performance of the implemented system with a standard deviation in the distance estimation of well below 1 cm in all cases, a total accuracy of at least 5 cm to 15 cm in an indoor environment depending on the concrete measurement scenario, as well as a good multipath resolution ability. The distance error has been found to be Gaussian distributed if optimal measurement conditions are given. The corresponding best-case standard deviation is 7 mm. A successful synchronization of two wireless units and distance measurements could be performed over a distance of more than 50 m. Due to its large signal bandwidth, the pulsed frequency modulated ultra-wideband system is robust against multipath interference and therefore a promising alternative to existing FMCW or UWB localization radars for indoor applications.

The realized demonstrator system shows the potential of the PFM-UWB technology for positioning solutions. However, there are some challenges to face to enable the transfer of the gained knowledge into a commercial product. Although a complete 3D positioning system was built up and was successfully evaluated in an industrial environment, there is still a lot of work to do. Aspects such as further global system optimization, form factor or power consumption have to be further considered and improved.

The results of this work give rise to many interesting future research topics. One possibility to realize a more efficient signal generator is to use a switched injection locked oscillator instead of chopping a continuous signal. Vossiek et al. developed such a signal synthesizer and showed the feasibility for ranging and positioning systems [120–122]. A switched injection-locked oscillator transponder can produce an approximately phase coherent response to an interrogating signal and, consequently, allows for transponder systems and precise distance measurement between the two wireless units.

Furthermore the antennas propose a field for future research and optimization. Although the presented system performs very good with the used antennas, a more specifically antenna in terms of directivity and antenna gain could improve the overall system performance. A second interesting topic is the implementation of an UWB antenna array to gain knowledge of the direction of the incoming signal. With such a system architecture, a combined TOF and AOA UWB position estimation would be possible.

Furthermore Goetz et al. studied the effects of short multipath propagation and presented simulation results for an applicable mitigation technique [123, 124]. To minimize distortions caused by multipath effects, a novel technique for mitigation of short multipath distortions is presented. It is based on the analysis of the distorted energy-spectral-density and processing of suitable characteristics by a neural network. The performance is investigated using the system parameters of the PFM-UWB prototype. The technique promises substantial improvements compared to the common maximum detection procedure in simulation and has to be verified by measurements.

Finally a chip integration of the proposed RF frontend would be an attractive way to reduce power consumption, the size, and the price of the measurement system. First research work for PFM-UWB chip integration is currently done at the University of Erlangen-Nuremberg. Compact and cheap units would open up new application fields like mobile personal information and navigation systems, location based services or RFID-like asset management services for the mass market.

APPENDIX A

UWB Emission Measurement

For compliance tests, the UWB bandwidth, the peak power and the average power have to be estimated. For ultra-wideband systems, it seems that there are two different philosophies to consider the measurement techniques of the radio signals: the main purpose of the first technique is to achieve a very accurate measurement result of the physical quantities, these measurements use very expensive instruments with high accuracy and precision and necessitate very time consuming measurement procedures. The second group of measurements consists of the regulatory or the compliance measurements. This second measurement technique uses off-the-shelf instruments and complete within a feasible time period. In [125] two approaches for measuring the emissions of an UWB system are introduced. The first one involves the measurement of the time domain characteristics of the signal. Via the discrete Fourier transform (DFT) or the FFT, these measurement results can be transfered into the frequency domain. This technique is suitable especially for measuring the peak power emission with wider RBW that cannot be achieved with a common spectrum analyzer. Using an oscilloscope suffers from a relatively narrow dynamic range, preventing the measurement of low level emission outside the UWB band. The alternative approach is to measure the UWB spectral characteristics in the frequency domain using a spectrum analyzer. These measurement results are influenced by the sweep time, the measurement time window of the spectrum analyzer, as well as by the duty ratio and the pulse repetition frequency of the measured signal. Depending on which signal characteristic is to be measured and which measurement equipment is available, the according measurement approach has to be chosen. This chapter deals with the latter approach, which has been used to evaluate the compliance of the developed UWB system.

A.1   Determination of UWB Bandwidth

To determine the UWB signal bandwidth the measurement is made using a spectrum analyzer with a resolution bandwidth (RBW) of 1 MHz and a video bandwidth (VBW) of 3 MHZ. The analyzer is set to a maximum-hold trace mode and to use a peak detector. The frequency point with the highest radiated emission in the peak PSD is designated as f_M, the frequency bins below and above f_M, where the emission level drops by 10 dB, are designated f_l and f_u according to chapter 2.1.

A.2   Average Power Spectral Density

The average power spectral density is defined as the maximum EIRP within 1 MHz bandwidth averaged over a maximum of 1 ms. This measurement is done by using the root mean square (RMS) detector to indicate the average in the spectrum analyzer. Table A.1 depicts the complete measurement settings for the spectrum analyzer.

RBW	1 MHz
VBW	3 x RBW
Detector	RMS
Sweep time	(number of bins) x 1 ms
Average PSD	each bin

Table A.1: Instruments settings for measuring the UWB average power spectral density using a spectrum analyzer.

A.3   Peak Power Spectral Density

The peak power spectral density is defined as the maximum EIRP within 50 MHz bandwidth with the observation time window of 0.1 ms. The measurement is centered on the frequency at which the highest emission occurs. Since hardly any spectrum analyzer is able to measure with a RBW of 50 MHz, the measurement is done with a narrower RBW and the emission limit is converted to the appropriate RBW. The FCC has provided a formula for redefining the peak emission limit of 0 dBm/50MHz in terms of the used RBW:

$$\text{EIRP limit in dBm} = 20 \log_{10}(\text{RBW used in MHz}/50). \quad (A.1)$$

APPENDIX B

PLL System Simulation

To make predictions about the RF signal quality and characteristics it is necessary to develop an accurate simulation model of the RF synthesizer. In this chapter the used PLL simulation models are described in detail and how they are implemented in Matlab Simulink.

B.1   Small-Signal Model

In the following section the small-signal model for the simulation of a PLL is derived as it is used for the system simulations. It is a linear model, which is often found in the literature and that can very well provide a fundamental understanding of the dynamics.

Transition to a Time-Dependent Phase

The small-signal model of the phase locked loop uses phases as input and output variables. Therefore, the transition from a sinusoidal time signal into a corresponding phase signal has to be derived. For reasons of clarity the time signal $U_{comp}(t)$ is used as reference signal. It can be represented as:

$$U_{comp}(t) = \widehat{U}_{comp} \cos\left(2\pi f_{comp}t + \phi_{comp}(t)\right). \tag{B.1}$$

The output signal is the time signal $U_{out}(t)$ with

$$U_{out}() = \widehat{U}_{out} \cos\left(2\pi f_{out}t + \phi_{out}(t)\right). \tag{B.2}$$

The amplitudes are not significant and are neglected below. The high-frequency signals are not modeled in the small-signal model. The frequencies f_{comp} and f_{out} are expressed by the appropriate divide ratios $1/R$ and $1/N$. Therefore in the model, only the time varying phases ϕ_{comp} and ϕ_{out} are variables of interest.

Block Diagram and Transfer Function

The small-signal model is a linear, time invariant model, which results from a linearization of the system in a given operating point. This operating point is the locked state, where it is assumed that the frequencies and phases at the input and output of the PFD at the starting point are the same size and additionally all state variables are stable. All system variables (such as the input and output phases or the control voltage of the VCO) are to be interpreted as small-signal parameters, which means as a deviation from the stable operating point. In Fig. B.1 the block diagram of the small-signal model according to [113] is depicted. In reality both the PFD and the the loop filter have an inverting characteristic.

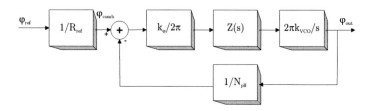

Figure B.1: Block diagram of the small-signal model of the closed loop of the PLL.

These inversions cancel each other and can be neglected in this model any further. The loop filter impedance can therefore be expressed as

$$Z\left(s\right) = \frac{1 + s\,T2}{s\left(A3\,s^3 + A2\,s^2 + A1\,s + A0\right)}.$$ (B.3)

By a suitable choice of sign in the PFD it is ensured that positive values for $\delta\phi$ results in a positive current at the output of the PFD. The transfer function is defined by

$$G\left(s\right) = \frac{k_\phi k_{VCO} Z\left(s\right)}{s}$$ (B.4)

and

$$H = \frac{1}{N_{pll}}.$$ (B.5)

Hence, the transfer function of the closed loop is given with

$$CL\left(s\right) = \frac{G\left(s\right)}{1 + G\left(s\right)H} = \frac{kN_{pll}\left(1 + s\,T2\right)}{s^5 A3 + s^4 A2 + s^3 A1 + s^2 A0 + s\,k\,T2 + k},$$ (B.6)

with

$$k = \frac{k_\phi k_{VCO}}{N_{pll}}.$$ (B.7)

This transfer function is a phase transfer function since it relates ϕ_{comp} and ϕ_{out} according to the following equation:

$$\phi_{out}\left(s\right) = CL\left(s\right)\phi_{comp}\left(s\right).$$ (B.8)

By defining the basic transfer function of $CL(s)$, it is now possible to make a conversion to other selected input variables. The following table summarizes the modified transfer functions. Obviously almost all transfer functions (reference source, R divider, N divider, PFD), have a common factor that significantly influences their response and have a low-pass characteristic. The functions differ only in their constant prefactors. The transfer function excited by the VCO is fundamentally different from the other transfer functions and has a high-pass characteristic. These different characteristics have great influence on the spurious and the phase noise. In Fig. B.2 the implemented small-signal model in the Matlab Simulink environment is shown.

Excitation by	Transfer function
Reference source	$\frac{1}{R}\frac{G(s)}{1+G(s)H}$
R divider	$\frac{G(s)}{1+G(s)H}$
N divider	$\frac{G(s)}{1+G(s)H}$
PFD	$\frac{1}{k_\phi}\frac{G(s)}{1+G(s)H}$
VCO	$\frac{1}{1+G(s)H}$

Table B.1: Conversion of the transfer function of $CL(s)$ to other input variables.

Definition of Bandwidth and Phase Margin

For the definition of the bandwidth f_c and the phase margin Φ the open-loop transfer function $G(s)H$ is used. The bandwidth f_c is that frequency at which the the gain frequency response is one or $0\,\text{dB}$:

$$|G(j2\pi f_c)H| = 1. \tag{B.9}$$

This system bandwidth is often called cutoff frequency but must not be mixed up with the cutoff frequency of the loop filter. The phase margin gives the difference of the phase frequency response at the frequency f_c to $180°$:

$$\Phi = 180° + \angle(G(j2\pi f_c)H). \tag{B.10}$$

It is a measure for the stability of the system and should therefore be dimensioned sufficiently large. The phase response also exhibits a local maximum at the frequency f_c. This will improve the stability and is achieved by appropriate choice of the filter components.

Properties of the Small-Signal Model

The small-signal model is very common in the literature and offers a number of advantages:

- It is a linear time invariant system and allows an analytic calculation of the transmission behavior.

- The settling time, stability, spurious suppression and phase noise can be calculated and simulated on the basis of the transfer functions.

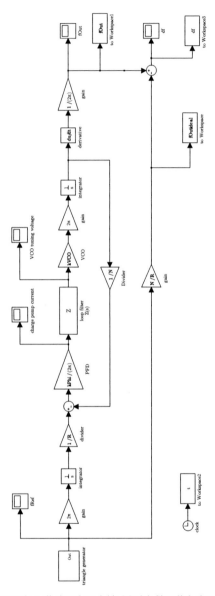

Figure B.2: Implemented small-signal model in Matlab Simulink simulation environment.

- High-frequency components are not included in the model. This has a very positive effect on the simulation time, since a relatively large time step is possible.

The disadvantages of this model are:

- There are assumptions for the model that might be not adequate. In particular, the output current of the PFD is assumed to be continuously and it is also assumed that the characteristic of the PFD is linear, which is not the case for the applied architecture. The effect can however be neglected if the bandwidth f_c is sufficiently small compared to the frequency of the current pulses.

- The model assumes that the phase locked loop is locked. All state variables are to be interpreted as deviations from this operating point. The model works better, the smaller the deviations are.

B.2   Large-Signal Model

The large-signal model tries to map the real processes in the phase locked loop more accurate. The time functions are completely included in the model, i.e. there is no transfer to the phase domain as it is in the small-signal model. Both the input functions and the output functions are sinusoidal. An analytical study of this model is no longer possible. The simulations are therefore done purely numerically using Matlab Simulink. The implemented large-signal model in the simulation environment is shown in Fig. B.3.

Modeling of the Phase Frequency Detector

The PFD is modeled as a state machine. This results in a nonlinear system, which analytically can not be investigated any further. The detector is built in Matlab Simulink in accordance with its real structure. In Fig. B.4 the block diagram of the implemented PFD model is depicted. The comparison blocks at the input convert the sinusoidal input signals into square wave signals. The D-flipflop and the NAND gates are available as prefabricated blocks.

Modeling of the VCO

In the large-signal model the VCO needs to provide a sinusoidal output signal, whose instantaneous frequency can be adjusted by the tuning voltage U_{VCO}. This is implemented according to Fig. B.5. The instantaneous frequency is generated at the output of the adder by the gain k_{VCO} and the constant f_0 By a multiplication by 2π angular frequencies are obtained, which are transformed into phase values by integration. These phases will be handed over as argument to the sine functions. Thus the sinusoidal time functions are created with the desired frequency. The subsequent gain stages allow a choice of the amplitude of the sine functions. These amplitudes are chosen constant equal to one. The $1/N_{pll}$ frequency divider is already integrated as a part in this sub-model.

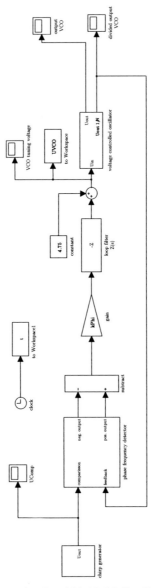

Figure B.3: Implemented large-signal model in Matlab Simulink simulation environment.

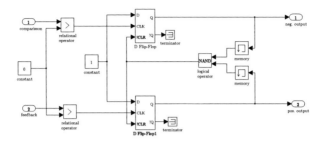

Figure B.4: Large-signal model of the PFD in Matlab Simulink simulation environment.

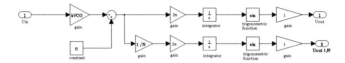

Figure B.5: Large-signal model of the VCO in Matlab Simulink simulation environment.

Modelling of the Reference Source

The reference source in this model has also to provide a sinusoidal output signal and additionally the triangular frequency modulation should be considered and implemented from the start. The generation of the output signal is analogous to the modeling of the VCO. However the instantaneous frequency here, is obtained with a different arrangement. The signal generator generates a bipolar square wave signal, which results in combination with the integrator, in a triangular signal. By adding a constant value the resulting instantaneous frequency is again obtained at the output of the adder. The implementation of the sub-model in Matlab Simulink is depicted in Fig. B.6.

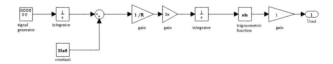

Figure B.6: Large-signal model of the reference source in Matlab Simulink simulation environment.

Properties of the Large-Signal Model

The large-signal model offers several advantages:

- The real structure of the PFD is considered. This makes it possible to study the influence of the spurious in detail. The current limitations of the small-signal model like suitably selected frequencies and bandwidths can therefore be omitted.

- The values are absolute values and do not depend on a chosen operating point.

- The model can be expanded gradually as needed. It is conceivable, for example, to implement a nonlinear control characteristic of the VCO using a table of values or to model the behavior of the frequency dividers more precisely.

The disadvantage of this model is:

- Because of the usage of the real signal forms the simulation time increases significantly. The sinusoidal signals have to be modeled accurately and this means that very small time steps have to be chosen.

APPENDIX C

Pulse Generator Architectures

For the PFM-UWB system, the generation of very narrow pulses are critical for the system performance. Such pulses can be generated by a variety of techniques. The aim of these circuits is to generate a pulse as short as possible and a certain periodicity. The time duration of the generated pulses is usually between 0.1 ns to 10 ns. In the following three pulse generating methods will be discussed: the avalanche transistor principle, step recovery diode pulse generating, and the use of a self-resetting D flip-flop.

D Flip-Flop

The most simple way to generate a narrow pulse is a D flip-flop with a feed back loop to the reset input. Fig. C.1 depicts the logic circuit. The input D of the flip-flop is always *high*, which results in a *high* impulse at the output Q with every clock edge. The feed back loop from $\overline{Q}$ to the reset input R resets the flip-flop immediately after setting. With this architecture and modern fast D flip-flops like the *74AHCT74*, pulse durations as low as 1.5 ns can be reached [126].

Step Recovery Diodes

Another common way to generate narrow pulses for UWB applications is the use of a step recovery diode (SRD). In Fig. C.2 a simple pulse generator according to [126, 127] is shown. In that circuit, the SRD is used as a charge controlled switch. With no signal at the input of that circuit, a charge is inserted onto the diode due to the forward biasing and therefore appears as a low impedance. If the charge is completely removed, the SRD switches very fast from a low impedance to a high impedance state. This feature can be used to generate fast pulses. In Fig. C.2 the first SRD is used to generate a fast rise time pulse, while the second SRD is responsible for a steep fall time by fast changing from low to high impedance state. With this circuit it is possible to generate pulses with a duration as low as 400 ps.

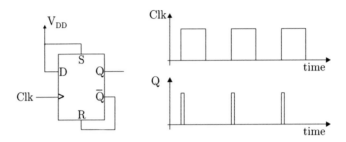

Figure C.1: TTL impulse generator using a D flip-flop with feed back loop to the reset pin.

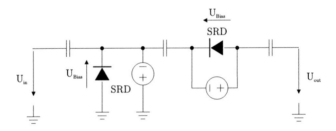

Figure C.2: Basic circuit of a SRD pulse generator. The diodes are used as a pulse sharpening circuit for raising and falling edges. With this circuit sub-nanosecond pulses can be generated.

Avalanche Transistor

With the use of so-called avalanche transistor circuits it is possible to achieve very short pulses with high pulse amplitudes [128]. Fig. C.3 shows the principle setup of an avalanche transistor pulse generator circuit. It consists of a bipolar transistor, whose collector-base transition is operating in the avalanche or breakdown mode. The collector is supplied over the resistor R_C by the voltage V_0, which is in the range of about 100 V to 300 V, depending on the breakdown voltage of the transistor. V_0 should be chosen in a way that the emitter voltage is near the transistor's avalanche breakdown voltage. V_0 also charges the transmission line, which serves as a charge storage for the to be generated pulse. If the trigger at the base puts the transistor in a break down mode, the whole charge of the transmission line will be discharged over the emitter resistor R_E very fast. These kinds of circuits can achieve slew rates in the range of 100 V/ns up to 1000 V/ns. After the pulse generation, the transmission line has to be recharged for the next pulse. This can take a certain time and limits the pulse repetition rate to a certain maximum. Other disadvantages, like the need for very high voltages or the weak repeatability of the generated pulse shapes made this pulse generation technique unpractical for the PFM-UWB system. Nevertheless, a detailed description and analysis of this architecture can be found in [126].

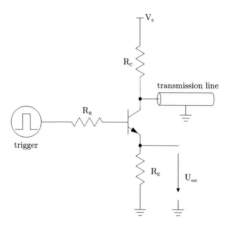

Figure C.3: Exemplary avalanche transistor circuit for pulse generation.

Bibliography

[1] Federal Communications Commission, "Part 15 - Radio Frequency Devices," july 2008.

[2] Wikipedia, [online] http://en.wikipedia.org/wiki/Positioning-system.

[3] S. Ingram, D. Harmer, and M. Quinlan, "Ultrawideband indoor positioning systems and their use in emergencies," in *Position Location and Navigation Symposium, PLANS*, 2004, pp. 706 – 715.

[4] Idtechex, [online] http://www.idtechex.com.

[5] Ubisense, [online] http://www.ubisense.net.

[6] Ubisense, [Datasheet] "Series 7000 Sensor", [online] http://www.ubisense.net/en/resources/fact-sheets.

[7] Ubisense, [Datasheet] "Series 7000 Compact Tag", [online] http://www.ubisense.net/en/resources/fact-sheets.

[8] Ubisense, [Datasheet] "Series 7000 Slim Tag", [online] http://www.ubisense.net/en/resources/fact-sheets.

[9] Time Domain, [online] http://www.timedomain.com.

[10] D. Kelly, S. Reinhardt, R. Stanley, and M. Einhorn, "PulsON second generation timing chip: enabling UWB through precise timing," in *Digest of Papers IEEE Conference on Ultra Wideband Systems and Technologies*, 2002, pp. 117 – 121.

[11] A. Petroff, R. Reinhardt, R. Stanley, and B. Beeler, "Pulson p200 UWB radio: simulation and performance results," in *Digest of Papers IEEE Conference on Ultra Wideband Systems and Technologies*, 2003, pp. 344 – 348.

[12] Time Domain, [Datasheet] "Plus System", [online] http://www.timedomain.com/datasheets/plus-system.php.

111

[13] Zebra Enterprise Solutions, [online] http://zes.zebra.com.

[14] R. Fontana and S. Gunderson, "Ultra-wideband precision asset location system," in *Digest of Papers IEEE Conference on Ultra Wideband Systems and Technologies*, 2002, pp. 147 – 150.

[15] R. Fontana, E. Richley, and J. Barney, "Commercialization of an ultra wideband precision asset location system," in *Digest of Papers IEEE Conference on Ultra Wideband Systems and Technologies*, 2003, pp. 369 – 373.

[16] R. Fontana, "Recent system applications of short-pulse ultra-wideband (uwb) technology," *IEEE Transactions on Microwave Theory and Techniques*, vol. 52, no. 9, pp. 2087 – 2104, sept. 2004.

[17] Zebra Enterprise Solutions, [Datasheet] "Sapphire Technology", [online] http://zes.zebra.com/products/rtls/tags-and-call-tags/sapphire.jsp.

[18] Zebra Enterprise Solutions, [Datasheet] "Sapphire DART Hub and Receiver", [online] http://zes.zebra.com/products/rtls/tags-and-call-tags/sapphire.jsp.

[19] Zebra Enterprise Solutions, [Datasheet] "Sapphire Tags", [online] http://zes.zebra.com/products/rtls/tags-and-call-tags/sapphire.jsp.

[20] P. Cheong, A. Rabbachin, J.-P. Montillet, K. Yu, and I. Oppermann, "Synchronization, toa and position estimation for low-complexity ldr uwb devices," in *IEEE International Conference on Ultra-Wideband, ICU*, 2005, pp. 480 – 484.

[21] J. Romme and B. Kull, "A low datarate localization system," in *Proceedings of UWB4SN 2005: workshop on UWB for Sensor Networks*, nov. 2005.

[22] A. Fujii, H. Sekiguchi, M. Asai, S. Kurashima, H. Ochiai, and R. Kohno, "Impulse radio uwb positioning system," in *IEEE Radio and Wireless Symposium*, 2007, pp. 55 – 58.

[23] C. Zhang, M. Kuhn, B. Merkl, M. Mahfouz, and A. Fathy, "Development of an UWB indoor 3D positioning radar with millimeter accuracy," in *IEEE MTT-S International Microwave Symposium Digest*, 2006, pp. 106 – 109.

[24] M. R. Mahfouz, C. Zhang, B. C. Merkl, M. J. Kuhn, and A. E. Fathy, "Investigation of high-accuracy indoor 3-d positioning using UWB technology," *IEEE Transactions on Microwave Theory and Techniques*, vol. 56, no. 6, pp. 1316 – 1330, june 2008.

[25] C. Zhang, M. Kuhn, B. Merkl, A. Fathy, and M. Mahfouz, "Real-time noncoherent uwb positioning radar with millimeter range accuracy: Theory and experiment," *IEEE Transactions on Microwave Theory and Techniques*, vol. 58, no. 1, pp. 9 – 20, jan. 2010.

[26] Z. Sahinoglu, S. Gezici, and I. Güvenc, *Ultra-wideband Positioning Systems: Theoretical Limits, Ranging Algorithms, and Protocols*, illustrated edition ed.   Cambridge University Press, oct. 2008.

[27] I. Oppermann, M. Haemaelaeinen, and J. Iinatti, Eds., *UWB: Theory and Applications.* Wiley, nov. 2004.

[28] J. D. Taylor, Ed., *Ultra-wideband Radar Technology*, 1st ed. CRC Press, sept. 2000.

[29] H. Hertz, "Über sehr schnelle elektrische Schwingungen," *Annalen der Physik und Chemie*, vol. 267, no. 7, pp. 421 – 448, 1887.

[30] M. Ghavami, L. Michael, and R. Kohno, *Ultra Wideband Signals and Systems in Communication Engineering*, 2nd ed. Wiley, mar. 2007.

[31] P. Runkle, J. McCorkle, T. Miller, and M. Welborn, "Ds-cdma: the modulation technology of choice for uwb communications," in *Digest of Papers IEEE Conference on Ultra Wideband Systems and Technologies*, 2003, pp. 364 – 368.

[32] The Commission of the European Communities, "Commission Decision of 21 February 2007 on allowing the use of the radio spectrum for equipment using ultra-wideband technology in a harmonised manner in the Community," *Official Journal of the European Union*, vol. 2007/131/EC, feb. 2007. [Online]. Available: http://eur-lex.europa.eu/LexUriServ/LexUriServ.do?uri=OJ: L:2007:055:0033:0036:EN:PDF

[33] Northern Digital Inc., [online] http://www.ndigital.com.

[34] Cybernet Systems Corporation, [online] http://www.cybernet.com.

[35] PMDTechnologies GmbH, [online] http://www.pmdtec.com.

[36] Sonitor Technologies, [online] http://www.sonitor.com.

[37] N. B. Priyantha, A. Chakraborty, and H. Balakrishnan, "The Cricket Location-Support System," in *MobiCom '00: Proceedings of the 6th annual international conference on mobile computing and networking.* New York, NY, USA: ACM, 2000, pp. 32 – 43.

[38] B. Krach and P. Robertson, "Integration of foot-mounted inertial sensors into a bayesian location estimation framework," in *5th Workshop on Positioning, Navigation and Communication, WPNC*, mar. 2008, pp. 55 – 61.

[39] Y. Chen and H. Kobayashi, "Signal Strength Based Indoor Geolocation," in *Proceedings of IEEE International Conference on Communications*, vol. 1, 2002, pp. 436 – 439.

[40] S.-H. Fang and T.-N. Lin, "Accurate wlan indoor localization based on rss fluctuations modeling," in *IEEE International Symposium on Intelligent Signal Processing, 2009. WISP*, aug. 2009, pp. 27 – 30.

[41] G. Wassi, C. Despins, D. Grenier, and C. Nerguizian, "Indoor location using received signal strength of ieee 802.11b access point," in *Canadian Conference on Electrical and Computer Engineering*, may 2005, pp. 1367 – 1370.

[42] Ekahau, [online] http://www.ekahau.com.

[43] P. Pathirana, A. Bishop, and A. Savkin, "Localization of mobile transmitters by means of linear state estimation using rss measurements," in *10th International Conference on Control, Automation, Robotics and Vision, ICARCV*, dec. 2008, pp. 210 – 213.

[44] J. Shirahama and T. Ohtsuki, "RSS-based localization in environments with different path loss exponent for each link," in *IEEE Vehicular Technology Conference, VTC 08 Spring*, may 2008, pp. 1509 – 1513.

[45] B. Turgut and R. Martin, "Localization for indoor wireless networks using minimum intersection areas of iso-rss lines," in *32nd IEEE Conference on Local Computer Networks, LCN*, oct. 2007, pp. 962 – 972.

[46] W. Wang and Q. Zhu, "RSS-based monte carlo localisation for mobile sensor networks," *IET Journal on Communications*, vol. 2, no. 5, pp. 673 – 681, may 2008.

[47] K. Whitehouse, C. Karlof, and D. E. Culler, "A Practical Evaluation of Radio Signal Strength for Ranging-based Localization," *Mobile Computing and Communications Review*, vol. 11, no. 1, pp. 41 – 52, 2007.

[48] S. Gezici, "A Survey on Wireless Position Estimation," *Wireless Personal Communications*, vol. 44, no. 3, pp. 263 – 282, feb. 2008.

[49] Y. Luo and C. L. Law, "UWB indoor positioning based on uniform circular antenna array," in *11th IEEE Singapore International Conference on Communication Systems, ICCS*, nov. 2008, pp. 138 – 142.

[50] W. Kocanda and A. Rutkowski, "Direction finding device with quasi-circular antenna array," in *14th International Conference on Microwaves, Radar and Wireless Communications, MIKON*, vol. 1, 2002, pp. 67 – 70.

[51] Y. Luo and C. L. Law, "Angle-of-arrival estimation with array in a line-of-sight indoor UWB-IR," in *7th International Conference on Information, Communications and Signal Processing, ICICS*, dec. 2009, pp. 1 – 5.

[52] J. Reilly, J. Litva, and P. Bauman, "New angle-of-arrival estimator: comparative evaluation applied to the low-angle tracking radar problem," *IEEE Proceedings, Communications, Radar and Signal Processing*, vol. 135, no. 5, pp. 408 – 420, oct. 1988.

[53] V. Kezys, E. Vertatschitsch, T. Greenlay, and S. Haykin, "High-Resolution Techniques for Angle-of-Arrival Estimation," in *IEEE Military Communications Conference - Communications-Computers: Teamed for the 90's, MILCOM*, vol. 3, oct. 1986, pp. 41.3.1 – 41.3.6.

[54] M. Grice, J. Rodenkirch, A. Yakovlev, H. Hwang, Z. Aliyazicioglu, and A. Lee, "Direction of arrival estimation using advanced signal processing," in *3rd International Conference on Recent Advances in Space Technologies, RAST '07*, june 2007, pp. 515 – 522.

[55] D. Dardari, A. Conti, U. Ferner, A. Giorgetti, and M. Win, "Ranging with ultrawide bandwidth signals in multipath environments," *Proceedings of the IEEE*, vol. 97, no. 2, pp. 404 – 426, feb. 2009.

[56] M. Vossiek, L. Wiebking, P. Gulden, J. Wieghardt, C. Hoffmann, and P. Heide, "Wireless local positioning," *IEEE Microwave Magazine*, vol. 4, no. 4, pp. 77 – 86, dec. 2003.

[57] A. Stelzer, K. Pourvoyeur, and A. Fischer, "Concept and application of LPM - a novel 3-d local position measurement system," *IEEE Transactions on Microwave Theory and Techniques*, vol. 52, no. 12, pp. 2664 – 2669, dec. 2004.

[58] M. Vossiek, R. Roskosch, and P. Heide, "Precise 3-D object position tracking using fmcw radar," in *29th European Microwave Conference*, vol. 1, oct. 1999, pp. 234 – 237.

[59] J. Heidrich, D. Brenk, J. Essel, G. Fischer, R. Weigel, and S. Schwarzer, "Local positioning with passive UHF RFID transponders," in *IEEE MTT-S International Microwave Workshop on Wireless Sensing, Local Positioning, and RFID, IMWS*, 2009, pp. 1 – 4.

[60] S. Roehr, P. Gulden, and M. Vossiek, "Precise distance and velocity measurement for real time locating in multipath environments using a frequency-modulated continuous-wave secondary radar approach," *IEEE Transactions on Microwave Theory and Techniques*, vol. 56, no. 10, pp. 2329 – 2339, oct. 2008.

[61] T.-Y. Chen, C.-C. Chiu, and T.-C. Tu, "Mixing and Combining with AOA and TOA for the Enhanced Accuracy of Mobile Location," in *5th European Personal Mobile Communications Conference, (Conf. Publ. No. 492)*, apr. 2003, pp. 276 – 280.

[62] V. Zhang and A.-S. Wong, "Combined AOA and TOA NLOS localization with nonlinear programming in severe multipath environments," in *IEEE Wireless Communications and Networking Conference, WCNC*, apr. 2009, pp. 1 – 6.

[63] N. Iwakiri and T. Kobayashi, "Joint TOA and AOA estimation of UWB signal using time domain smoothing," in *2nd International Symposium on Wireless Pervasive Computing, ISWPC '07*, feb. 2007.

[64] A. Hatami and K. Pahlavan, "Hybrid TOA-RSS Based Localization Using Neural Networks," in *IEEE Global Telecommunications Conference, GLOBECOM '06*, dec. 2006, pp. 1 – 5.

[65] T. Mogi and T. Ohtsuki, "TOA localization using RSS weight with path loss exponents estimation in NLOS environments," in *14th Asia-Pacific Conference on Communications, APCC*, oct. 2008, pp. 1 – 5.

[66] N. Thomas, D. Cruickshank, and D. Laurenson, "Performance of a TDOA-AOA Hybrid Mobile Location System," in *Second International Conference on 3G Mobile Communication Technologies, (Conf. Publ. No. 477)*, 2001, pp. 216 – 220.

[67] C. Yang, Y. Huang, and X. Zhu, "Hybrid TDOA/AOA-method for indoor positioning systems," in *The Institution of Engineering and Technology Seminar on
Location Technologies*, dec. 2007, pp. 1 – 5.

[68] F. Althaus, F. Troesch, and A. Wittneben, "UWB geo-regioning in rich multipath
environment," in *IEEE 62nd Vehicular Technology Conference, VTC 05 Fall*, vol. 2,
sept. 2005, pp. 1001 – 1005.

[69] M. Triki, D. Slock, V. Rigal, and P. Francois, "Mobile terminal positioning via
power delay profile fingerprinting: Reproducible validation simulations," in *IEEE
64th Vehicular Technology Conference, VTC 06 Fall*, sept. 2006, pp. 1 – 5.

[70] C. Nerguizian, C. Despins, and S. Affes, "Geolocation in Mines with an Impulse
Response Fingerprinting Technique and Neural Networks," *IEEE Transactions on
Wireless Communications*, vol. 5, no. 3, pp. 603 – 611, mar. 2006.

[71] R. E. Kalman, "A New Approach to Linear Filtering and Prediction Problems,"
Transactions of the ASME - Journal of Basic Engineering, vol. 82 (Series D), pp.
35 – 45, 1969.

[72] A. Doucet, N. De Freitas, and N. Gordon, *Sequential Monte Carlo Methods in
Practice*. Springer, 2001.

[73] K. Spingarn, "Passive position location estimation using the extended kalman filter," *IEEE Transactions on Aerospace and Electronic Systems*, vol. AES-23, no. 4,
pp. 558 – 567, july 1987.

[74] T. Perala and R. Piche, "Robust extended kalman filtering in hybrid positioning
applications," in *4th Workshop on Positioning, Navigation and Communication,
WPNC '07*, sept. 2007, pp. 55 – 63.

[75] S. Ali-Loytty, T. Perala, V. Honkavirta, and R. Piche, "Fingerprint kalman filter in
indoor positioning applications," in *IEEE Control Applications, (CCA) Intelligent
Control, (ISIC)*, 2009, pp. 1678 – 1683.

[76] D. Jourdan, J. Deyst, J.J., M. Win, and N. Roy, "Monte Carlo Localization in Dense
Multipath Environments using UWB Ranging," in *IEEE International Conference
on Ultra-Wideband, ICU*, 2005, pp. 314 – 319.

[77] B. Denis, L. Ouvry, B. Uguen, and F. Tchoffo-Talom, "Advanced Bayesian Filtering Techniques for UWB Tracking Systems in Indoor Environments," in *IEEE
International Conference on Ultra-Wideband, ICU*, 2005, p. 6 pp.

[78] H. Urkowitz, C. Hauer, and J. Koval, "Generalized Resolution in Radar Systems,"
Proceedings of the IRE, vol. 50, no. 10, pp. 2093 – 2105, oct. 1962.

[79] Z. Sahinoglu, S. Gezici, and I. Güvenc, *Ultra-wideband Positioning Systems:
Theoretical Limits, Ranging Algorithms, and Protocols*, illustrated edition ed.
Cambridge University Press, oct. 2008. [Online]. Available: http://amazon.com/o/
ASIN/0521873096/

[80] A. F. Molisch, "Ultrawideband propagation channels-theory, measurement, and modeling," *IEEE Transactions on Vehicular Technology*, vol. 54, no. 5, pp. 1528 – 1545, sept. 2005.

[81] J. Karedal, S. Wyne, P. Almers, F. Tufvesson, and A. F. Molisch, "Statistical analysis of the UWB channel in an industrial environment," in *IEEE 60th Vehicular Technology Conference, VTC 04 Fall*, vol. 1, sept. 2004, pp. 81 – 85.

[82] A. F. Molisch, K. Balakrishnan, D. Cassioli, C.-C. Chong, S. Emami, A. Fort, J. Karedal, J. Kunisch, H. Schantz, and K. Siwiak, "A comprehensive model for ultrawideband propagation channels," in *IEEE Global Telecommunications Conference, GLOBECOM '05*, vol. 6, St. Louis, MO, dec. 2005.

[83] A. F. Molisch, D. Cassioli, C.-C. Chong, S. Emami, A. Fort, B. Kannan, J. Karedal, J. Kunisch, H. G. Schantz, K. Siwiak, and M. Z. Win, "A comprehensive standardized model for ultrawideband propagation channels," *IEEE Transactions on Antennas and Propagation*, vol. 54, pp. 3151 – 3166, nov. 2006.

[84] B. Kannan, C. Kim, X. Sun, L. Chiam, C. F., and Y. Chew, "UWB channel characterization in office environments," IEEE, Tech. Rep. IEEE 802.15-04-0439-00-004a, 2004.

[85] D. Cassioli, Ciccognani, and A. Durantini, "D3.1-UWB Channel Model Report," ULTRAWAVES, Tech. Rep. IST-2001-35189, 2003.

[86] A. Durantini, W. Ciccognani, and D. Cassioli, "UWB propagation measurements by pn-sequence channel sounding," in *IEEE International Conference on Communications*, vol. 6, jun. 2004, pp. 3414 – 3418.

[87] B. Kannan, C. Kim, X. Sun, L. Chiam, C. F., and Y. Chew, "UWB channel characterization in outdoor environments," IEEE, Tech. Rep. IEEE 802.15-04-0440-00-004a, 2004.

[88] J. Keignart, J.-B. Pierrot, N. Daniele, A. Alvarez, M. Lobeira, J. L. Garcia, G. Valera, and R. P. Torres, "Ultra-wideband concepts for ad hoc networks report on UWB basic transmission loss," U.C.A.N., Tech. Rep. IST-2001-32710, 2003.

[89] R. C. Qiu and I.-T. Lu, "Wideband wireless multipath channel modeling with path frequencydependence," in *IEEE International Conference on Communications, ICC 96, Conference Record, Converging Technologies for Tomorrow's Applications*, vol. 1, Dallas, TX, USA, june 1996, pp. 277 – 281.

[90] R. Qiu and I.-T. Lu, "Multipath resolving with frequency dependence for wideband wireless channel modeling," *IEEE Transactions on Vehicular Technology*, vol. 48, no. 1, pp. 273 – 285, jan. 1999.

[91] J. Kunisch and J. Pamp, "Measurement results and modeling aspects for the UWB radio channel," in *Digest of Papers IEEE Conference on Ultra Wideband Systems and Technologies*, may 2002, pp. 19 – 23.

[92] P. Bello, "Characterization of randomly time-variant linear channels," *IEEE Transactions on Communications Systems*, vol. 11, no. 4, pp. 360 – 393, dec. 1963.

[93] C.-C. Chong, Y. Kim, and S.-S. Lee, "A modified S-V clustering channel model for the UWB indoor residential environment," in *IEEE 61st Vehicular Technology Conference, VTC 05 Spring*, vol. 1, may 2005, pp. 58 – 62.

[94] A. Saleh and R. Valenzuela, "A statistical model for indoor multipath propagation," *IEEE Journal on Selected Areas in Communications*, vol. 5, no. 2, pp. 128 – 137, feb. 1987.

[95] IEEE, "802.15.3c channel model final report," 2007. [Online]. Available: http://www.ieee802.org/15/pub/TG4a.html

[96] S. Venkatesh, J. Ibrahim, and R. M. Buehrer, "A new 2-cluster model for indoor UWB channel measurements," in *IEEE International Symposium Antennas and Propagation Society*, vol. 1, june 2004, pp. 946 – 949.

[97] D. Cassioli, A. Durantini, and W. Ciccognani, "The role of path loss on the selection of the operating bands of UWB systems," in *15th IEEE International Symposium on Personal, Indoor and Mobile Radio Communications, PIMRC*, vol. 4, sept. 2004, pp. 2787 – 2791.

[98] D. Cassioli and A. Durantini, "A time-domain propagation model of the UWB indoor channel in the FCC-compliant band 3.6 - 6 GHz based on pn-sequence channel measurements," in *IEEE 59th Vehicular Technology Conference, VTC 04 Spring*, vol. 1, may 2004, pp. 213 – 217.

[99] D. Cassioli, M. Z. Win, and A. F. Molisch, "The Ultra-Wide Bandwidth Indoor Channel: from Statistical Model to Simulations," *IEEE Journal on Selected Areas in Communications*, vol. 20, no. 6, pp. 1247 – 1257, aug. 2002.

[100] SkyCross, [Datasheet] "Ultra Wideband Antenna SMT-3TO10M-A", [online] http://www.skycross.com/Products/PDFs/SMT-3TO10M-A.pdf.

[101] B. Waldmann, R. Weigel, and P. Gulden, "Method for high precision local positioning radar using an ultra wideband technique," in *Proc. IEEE MTT-S International Microwave Symposium Digest*, june 2008, pp. 117 – 120.

[102] B. Waldmann, R. Weigel, P. Gulden, and M. Vossiek, "Pulsed frequency modulation techniques for high-precision ultra wideband ranging and positioning," in *Proc. IEEE International Conference on Ultra-Wideband ICUWB 2008*, vol. 2, sept. 2008, pp. 133 – 136.

[103] B. Waldmann, P. Gulden, M. Vossiek, and R. Weigel, "A pulsed frequency modulated ultra wideband technique for indoor positioning systems," *Frequenz*, vol. 62, no. 7-8, pp. 195 – 198, 2008.

[104] R. Mosshammer, B. Waldmann, R. Eickhoff, R. Weigel, and M. Huemer, "A comparison of channel access concepts for high-precision local positioning," in *6th Workshop on Positioning, Navigation and Communication, 2009. WPNC 2009.*, mar. 2009, pp. 37 – 41.

[105] Symeo GmbH, [online] http://www.symeo.com.

[106] S. Roehr, P. Gulden, and M. Vossiek, "Novel secondary radar for precise distance and velocity measurement in multipath environments," in *Proceedings of the European Microwave Conference*, oct. 2007, pp. 1461 – 1464.

[107] S. Roehr, P. Gulden, and M. Vossiek, "Novel secondary radar for precise distance and velocity measurement in multipath environments," in *Proceedings of the European Radar Conference EuRAD*, oct. 2007, pp. 182 – 185.

[108] S. Roehr, P. Gulden, and M. Vossiek, "Method for high precision clock synchronization in wireless systems with application to radio navigation," in *IEEE Radio and Wireless Symposium*, jan. 2007, pp. 551 – 554.

[109] S. Roehr, M. Vossiek, and P. Gulden, "Method for high precision radar distance measurement and synchronization of wireless units," in *IEEE/MTT-S International Microwave Symposium*, june 2007, pp. 1315 – 1318.

[110] S. Roehr, *System-Theoretic Analysis and Optimization of a Novel Secondary Radar Concept for Precise Distance and Velocity Measurement*. Logos Berlin, may 2010.

[111] D. C. von Grünigen, *Digitale Signalverarbeitung: mit einer Einführung in die kontinuierlichen Signale und Systeme.*, 4th ed. Hanser Fachbuch, oct. 2008.

[112] D. D. Wentzloff and A. P. Chandrakasan, "A 3.1-10.6 GHz ultra-wideband pulse-shaping mixer," *Digest of Papers IEEE Radio Frequency integrated Circuits (RFIC) Symposium*, pp. 83 – 86, june 2005.

[113] D. Banerjee, *PLL Performance, Simulation and Design*, 4th ed. Dog Ear Publishing, 2006.

[114] R. Mosshammer, "Cross-Layer Simulation Analysis of a High-Precision Radiolocation System," Ph.D. dissertation, University of Erlangen-Nuremberg, 2010.

[115] Avago Technologies, [Datasheet] "AMMC-2008 DC - 50 GHz SPDT Switch", [online] http://www.avagotech.com/docs/AV02-126EN.

[116] H. Schantz, "Introduction to ultra-wideband antennas," in *IEEE Conference on Ultra Wideband Systems and Technologies*, nov. 2003, pp. 1 – 9.

[117] E. Pancera, C. Sturm, and W. Wiesbeck, "Small UWB coplanar monopole antenna design," in *The Second European Conference on Antennas and Propagation, EuCAP*, nov. 2007, pp. 1 – 3.

[118] H. Schantz, "Planar elliptical element ultra-wideband dipole antennas," vol. 3, 2002, p. 44.

[119] Elspec GmbH, [Datasheet] "LL2773-AF: Koaxiales HF-Kabel", [online] http://www.elspec.de.

[120] M. Vossiek and P. Gulden, "The switched injection-locked oscillator: A novel versatile concept for wireless transponder and localization systems," *IEEE Transactions on Microwave Theory and Techniques*, vol. 56, no. 4, pp. 859 – 866, apr. 2008.

[121] M. Vossiek, T. Schafer, and D. Becker, "Regenerative backscatter transponder using the switched injection-locked oscillator concept," in *IEEE MTT-S International Microwave Symposium Digest*, 2008, pp. 571 – 574.

[122] T. Schafer, F. Kirsch, and M. Vossiek, "A 13.56 MHz localization system utilizing a switched injection locked transponder," in *IEEE International Conference on Microwaves, Communications, Antennas and Electronics Systems, COMCAS*, 2009, pp. 1 – 4.

[123] B. Waldmann, A. Goetz, and R. Weigel, "An ultra wideband positioning system enhanced by a short multipath mitigation technique," in *IEEE MTT-S International Microwave Workshop on Wireless Sensing, Local Positioning, and RFID, IMWS*, 2009, pp. 1 – 4.

[124] A. Goetz, B. Waldmann, and R. Weigel, "Short multipath mitigation technique using feedforward neural networks," in *German Microwave Conference*, 2009, pp. 1 – 4.

[125] J.-I. Takada, S. Ishigami, J. Nakada, E. Nakagawa, M. Uchino, and T. Yasui, "Measurement techniques of emissions from ultra wideband devices," *IEICE Trans. Fundam. Electron. Commun. Comput. Sci.*, vol. E88-A, no. 9, pp. 2252 – 2263, 2005.

[126] T. Buchegger, "Design and Simulation of Ultra Wideband Pulse-Based Communication Systems," Ph.D. dissertation, Johannes Kepler University Linz, oct. 2005.

[127] G. Ossberger, "High-Resolution Pulse-Based Ultra-Wideband Ground Penetrating Radar System for Detection of Buried Anti-Personnel Mines," Ph.D. dissertation, Johannes Kepler University Linz, aug. 2005.

[128] P. Spirito, "On the trigger delay of avalanche transistors," *IEEE Journal of Solid-State Circuits*, vol. 9, no. 5, pp. 307 – 309, oct 1974.